OVERCOMING THE TWO CULTURES

Fernand Braudel Center Series
Edited by Immanuel Wallerstein

Overcoming the Two Cultures

Science versus the Humanities in the Modern World-System

Richard E. Lee and Immanuel Wallerstein,
COORDINATORS

WITH

Volkan Aytar
Ayşe Betül Çelik
Mauro Di Meglio
Mark Frezzo
Ho-fung Hung
Biray Kolluoğlu Kırlı
Agustín Lao-Montes
Eric Mielants
Boris Stremlin
Sunaryo
Norihisa Yamashita
Deniz Yükseker

Routledge
Taylor & Francis Group

LONDON AND NEW YORK

First published 2004 by Paradigm Publishers

Published 2016 by Routledge
2 Park Square, Milton Park, Abingdon, Oxon OX14 4RN
711 Third Avenue, New York, NY 10017, USA

Routledge is an imprint of the Taylor & Francis Group, an informa business

Library of Congress Cataloging-in-Publication Data

Overcoming the two cultures : science versus the humanities in the modern world-system / Richard E. Lee and Immanuel Wallerstein, coordinators ; with Volkan Aytar . . . [et al.].
  p. cm — (Fernand Braudel Center series)
  Includes bibliographical references and index.
  ISBN 1-59451-068-7 (hardcover : alk. paper) — ISBN 1-59451-069-5 (pbk. : alk. paper)
  1. Interdisciplinary approach to knowledge. 2. Culture. 3. Science and the humanities. I. Lee, Richard E., 1945-   II. Wallerstein, Immanuel Maurice, 1930- III. Aytar, Volkan. IV. Series.

HM651.O94 2004
001—dc22

                                                                    2004016382

Designed and Typeset by Straight Creek Bookmakers.

ISBN 13: 978-1-59451-068-7 (hbk)
ISBN 13: 978-1-59451-069-4 (pbk)

Contents

1

Introduction: The Two Cultures

Richard E. Lee and Immanuel Wallerstein

The "two cultures" is a phrase invented by C. P. Snow in 1959,[1] but the phenomenon he was describing is of course older. It is not, however, all that old, and that is the point of this book. In terms of human history, the idea that there are two cultures is relatively recent. Furthermore, whereas when Snow wrote his book, the concept seemed self-evident, which is why Snow's phrase caught on, its validity has come under increasing challenge since the 1960s. We shall try to explain the origin of the concept, its impact and pervasiveness, and the nature of the challenges to it in recent years. This will, we hope, enable us to assess the likelihood that this way of structuring knowledge will continue to prevail and, if it does not, what the alternatives are.

The word "culture" in the phrase "two cultures" refers to the fact that scholars do their research, writing, and teaching on the basis of underlying epistemological presuppositions, which they use but seldom expound. To say that there are two cultures (in the structures of knowledge) is to say that scholars tend to group themselves in two different, indeed often opposing, camps with regard to the set of epistemological presuppositions they employ and believe useful and/or correct to employ.

The idea that there are two cultures is a comparatively simple one. The problem posed by Snow, "by training . . . a scientist: by vocation . . . a writer" (1993: 1), was that the two cultures did not "understand" each other. He was the master of a college at Cambridge, where fellows from many different disciplines met at high table. Snow observed,

> Constantly I felt I was moving among two groups—comparable in intelligence, identical in race, not grossly different in social origin, earning about the same incomes, who had almost ceased to communicate at all, who in intellectual, moral and psychological climate had so little in common that instead of going from Burlington House or South Kensington to Chelsea, one might have crossed an ocean. (1993: 2)

Snow felt that such lack of mutual understanding among scholars was dangerous and a social loss, and his lectures were an attempt to bridge the gap by explaining each of the two cultures to the other. Snow was of course certain that there did exist two cultures and that they were quite different one from the other, and that in a sense they ought to be different. For Snow, the fact that there were two cultures represented an intellectual division of labor that was intelligible and justifiable. He was primarily (perhaps only) concerned with minimizing the internecine strife that had resulted from this scholarly reality. We are, by contrast, asking how the gulf between the two cultures was created and whether it can be overcome. Is it being overcome? Should it be overcome?

What does one mean by the two cultures? What even are their names? The answer is not so simple. One of the two cultures is usually called the scientific culture. The other has many names: the literary, the philosophical, the humanistic culture. We shall see that the fact that the latter has many names, while the name of the "scientific" culture is largely agreed upon, is not completely accidental. For Snow, as for many others, the two cultures are not symmetrical. They exist in a hierarchy of importance and/or merit, although which is higher has been a subject of debate.[2] And they are of different longevity: the humanistic culture is considered to be the older one, the "traditional" one, whereas the scientific culture is usually said to be the newer, more "modern" one.[3]

We shall be challenging this chronology as well. We shall be arguing that, prior to the existence of the modern world-system, the structures of knowledge were radically different in their epistemology from those of the modern world. We shall be arguing that, in the modern world, it was what eventually came to be called the scientific culture that emerged first and that the creation of the humanistic culture was in large part a consequence of the creation of the scientific culture.

It is important to start the analysis by thinking about the premodern world—not merely the premodern world in Europe, or what we now call the West, but the premodern world across the globe. There were of course many, many differences among the multiple civilizations, or historical systems, that existed prior to, say, AD 1500, but there was, it seems to us, a common feature to their structures of knowledge. This common feature was an inability to conceive that there were two distinctive ways of knowing, two epistemologies that one could employ in different domains.

In the premodern world, to know was to know. And while one could argue about what one knew, one did not really argue about how one knew. The specialists in knowledge, the intellectuals, did not know everything. But what they knew or claimed to know did not fit within any of the boxes called disciplines that we have created in modern times. The so-called disciplines in fact took quite a while to be established even within the modern world. They were not there in 1500 and were barely there in 1800. They are creations largely of the nineteenth century. And we are not really sure today if they will continue to be there, at least in the form we now know them, in 2100.

What is it that one knows? Three kinds of things: what is true, what is good, what is beautiful. The definitions of each of these abstract concepts constitute precisely what we mean by a civilization. The details of the definitions are of course debated, often quite fiercely, within the general framework of each so-called civilization. But one of the elements that integrate the historic civilizations has been that each believed that it could define and obtain knowledge about the true, the good, and the beautiful. And up to the modern world, no one seemed to think that these three objectives of knowledge were segmented, separable activities. Keats's line of poetry "Truth is beauty, beauty truth" would have resonated well in all the historic civilizations, although it seems romantic and quaint today, and already did so in nineteenth-century Europe when it was written.

The story we shall tell seems to us quite straightforward. The first part of this book concerns the historical construction and institutionalization of the two cultures. This, in effect, is the story of the construction of an altogether new and unique structure, or "whole," organized as a relational hierarchy of disciplinary "parts." However, we are not writing a history of ideas. We are asking which ideas were widely accepted or influential and received the sanction of becoming the basis of institutions of knowledge, such that they were passed on, perpetuated, and developed by successive generations of intellectuals, scholars, scientists, in direct relation with the material processes of the modern world-system.

Concepts are not constructed in an instant. There is often a long process of slow gestation. Once a concept is constructed, one can always locate forebears. There is scarcely ever a totally new idea. But this does not concern us. We are interested in when and how some concepts become widely accepted as the basis of ongoing structures of knowledge. We believe, as you shall see, that the scientific culture was growing in importance and autonomy from the beginning of the modern world-system, which we date at the second half of the fifteenth century, even if the scientific culture as we know it today was not fully institutionalized until the nineteenth century. We believe that as the "scientific" culture, the putative realm of "truth" and "facts," was separated out of what had been theretofore an all-encompassing "philosophy," and a "humanistic" culture devoted to "willing the good" was constructed at the opposite pole of the consolidating structures of knowledge, there came about a major reaction, largely after the French Revolution, to the growing pre-eminence of the sciences. It includes but is not limited to romanticism. Thus, in the nineteenth century, the humanistic culture, or perhaps we should say a humanistic counterculture, was constructed defensively against the imperialist claims of the scientific culture.

And finally, in response to developments in the real world of human social relations that challenged the dominant modes of understanding in both the sciences and the humanities, in the latter half of the nineteenth century a third institutional arena, that of the social sciences, began to emerge, one that, we shall argue, has always been caught between the contradictory pulls of the by then well-established two cultures. Because the pressures on this arena were so great, the resulting confusion and lack of clarity has been perhaps greatest in this field of knowledge activity.

It seems clear to us that in this nineteenth-century struggle the scientific culture was able to impose itself socially as the dominant culture in the world of knowledge of the modern world-system. The degree of its supremacy seems to have been a steady upward curve until it reached a high point in the period 1945–1970. And then the scientific culture began to come under a severe counterattack. Externally, this occurred because of a broad-based shift in the political economy of the world-system, for which the world revolution of 1968 stands as a marker and a symbol. Internally, this occurred as an outcome of the process of development within the sciences themselves. We shall try to trace seven major intellectual movements, primarily in the period from the 1960s to the end of the twentieth century. They are movements of quite different sorts. The first is that of complexity studies, a movement from within the scientific culture that has been challenging the formerly dominant epistemology. A second is the social studies of science, which is a movement that has been asking questions about how scientific knowledge is constructed. Then there are the movements we are calling "diversity movements": feminisms and movements organized around race and ethnicity in the West. These movements exist both within the structures of knowledge and within the general political arena. We shall also consider popular culture/cultural studies, a movement that explicitly contests the divide between the humanities and the social sciences. We shall also look at ecology/environmentalism, again a movement that operates within both the structures of knowledge and the general political arena and, in this case, contests the divide between the sciences and the social sciences.

We shall try to outline the issues that are being raised by each movement and the degree to which they have been able to institutionalize themselves in the world university system and other loci of the structures of knowledge. But we shall do more than this. We shall try to evaluate the degree to which their concerns have led them to question the validity of the concept of the two cultures and to see how far they have been willing to go in this questioning. And finally, we shall look at the so-called culture wars and science wars as arenas of conflict over the validity of the concept of the two cultures. We believe that this kind of overall historical and analytic view of the ways in which the concept of the two cultures has served as a central pillar of the geoculture of the modern world-system has not hitherto been attempted. We believe further that the questioning of this concept that occurred in the last third of the twentieth century is a major cultural event, inseparable from important transformations in the world-system and contributing in its turn to further transformations. We think we are living in a sort of cultural hurricane or earthquake and that it is crucial for everyone to be very clear what are the issues and what are the alternative possible outcomes.

In analyzing this cultural past, we are participating in the struggle to create our cultural future.

Notes

1. C. P. Snow, *The Two Cultures* (Cambridge: Cambridge University Press, Canto edition, 1993).

2. Snow himself makes clear his hierarchy: "Most people, wherever they are being given a chance, are rushing into the scientific revolution. To misunderstand this position is to misunderstand the present and the future. It simmers beneath the surface of world politics" (1993: 80).

3. "If the scientists have the future in their bones, then the traditional culture responds by wishing the future did not exist. It is the traditional culture, to an extent remarkably little diminished by the emergence of the scientific one, which manages the western world" (1993: 11). Note here that Snow speaks of the "western" world rather than of the "modern" world. This is an often tacit premise.

*

PART I

Historical Construction of the Two Cultures

2

Constructing Authority: The Rise of Science in the Modern World

Boris Stremlin

During the European Middle Ages, the production of knowledge generally conformed to the model outlined by St. Augustine of Hippo. For
Augustine, "[t]he uses of a knowledge of languages, of history, of gram-
mar and even logic" lay in their capacity "as aids to the study of the
Bible" (Southern 1953: 171). The Scholasticism that arose on this foundation in the twelfth and thirteenth centuries has been described as purely
logical and "extreme rationalism," a "routine study of Latin," which,
owing to the chronic shortage of books, placed a "great emphasis on
disputation and oratory" (Bowle 1970: 151–52). Despite the clumsiness
of its methods, in aiming to "study problems of universal significance,"
Scholasticism was animated by a wide-ranging curiosity (Southern 1953:
170). Its authority remained, nonetheless, indissolubly linked to a single
and unquestioned set of values.

Ultimately, Scholasticism was undermined both internally and externally. From within, scholars eventually found themselves in agreement
with those, like William of Ockham, who argued that since the power of
God was absolute, it did not need to be explained rationally and therefore there was no need of finding ways of combining religion with
worldly thought. From outside the Scholastic community, a variety of

so-called humanist movements sought initially a modest objective—to reestablish what they perceived as the original form of Latin. But in so doing, they implicitly rejected the method of the Schoolmen who argued that Latin was the sole written and liturgical language and that therefore all ancient texts retained an air of inviolability, fit for commentary but not to be repaired. Over time, the consequence was that

> the exacting philology [developed by the Humanists], which made it possible to restore the authentic form of an ancient work . . . could not be confined to the writings of pagan antiquity . . . . Thus, the humanist method called into question the Catholic Church's most fundamental traditions, by the very procedures it followed. (Mandrou 1979: 47–49)

By the sixteenth century, the humanistic method had come to be widely used for the purpose of biblical exegesis and as such stimulated the proliferation of religious radicalism, both Calvinist (Moeller 1972) and Jesuit (Olin 1994). Moreover, its use was no longer restricted to Latin but applied to a variety of texts in Greek and Hebrew, which not only resulted in an increase of the number of texts available to the literate public but also uncovered theological traditions that many believed antedated that of the Catholic Church (Yates 1964: chs. 1, 4). Once the humanists embraced oral vernaculars, they shook the institutional foundations of the medieval structures of knowledge. The appearance of new secular literary genres—poetry, tales, romances, drama, and more localized conceptions of history—fostered the rise of new networks of readers located within circumscribed linguistic regions, no longer extending over the whole of Latin Christendom (Mandrou 1979: 130–31, 309–11).

Unlike the university Schoolmen, who were all technically representatives of the ecclesiastical estate and whose collective mission was to codify orthodox doctrine (French & Cunningham 1996) and teach it to all worthy students, the humanists shared no common social or professional ethic. Drawn to the best libraries, academies, observatories, collections of curiosities, and botanical gardens, many of them entered state service as ministers, diplomats, or artisans and came to resemble courtiers in their social outlook (Moran 1991; Biagioli 1992). Some became teachers at the new humanist colleges founded by princes or by the Jesuits, or at the few (notably Italian) universities that opened up to humanist learning (Schmitt 1984: chs. 14, 15). Other literati congregated around the new printing houses, which offered employment to translators, indexers, abridgers, and illustrators (Eisenstein 1983: 61–63). Some still depended on the mutual support offered by the informal network known as the Republic of Letters (Mandrou 1979: 55–57). Their lack of cohesion often elicited feelings of social isolation and contributed to the development of empiricism and the modern "mistrust of the social aspects of knowledge making" (Shapin 1996: 69).

As the centers of authority proliferated, the substantive worldview and epistemology of the Scholastics, formerly sacrosanct, also came under scrutiny. The natural philosophy developed by the Schoolmen and friars in the twelfth and thirteenth centuries "had centrally to do with God, with man's relation to Him and man's attempt to lead the good life as a Christian philosopher" (French & Cunningham 1996: 92). The geocentric Ptolemaic model incorporated into the Scholastic corpus reinforced the biblical worldview of Earth as the center of the divine drama, in which it was for the sake of humanity that the universe had been created. Although humanists accepted the authority of the ancients and the church fathers, new ways of conceptualizing the world were already emerging by the last quarter of the thirteenth century. A stress on visual representation, the homogeneity of space and time, and the apprehension of the universe through the medium of quantifiable units was beginning to bear on the practice of painting, music, business, and technology (Crosby 1997). There was also a growing concern for the practical uses of mathematics, a sign of the strengthened opposition to the subordination of practical knowledge to religious and contemplative knowledge (Høyrup 1994: 166–71).

Herein lies the germ of the structural differentiation, the separation of authoritative knowledge from social values unique in the history of human knowledge production, that has come to be known as the "two cultures." It achieved an initial institutional status, to be progressively deepened, and a range of social consensus that enabled it to become broadly influential during the period of the transition from feudalism to capitalism. As Mary Poovey (1998) has argued, the "modern fact" emerged as the primary epistemological unit of valid knowledge and cultural authority. The virtues of "balance" attributed to the system of double-entry bookkeeping legitimated profit. Discriminating between profit and usury unleashed the endless accumulation of capital. Both the specifics of commerce and their generalization within a system that underwrote the individual creditworthiness of merchants and their credibility as a group could be characterized as "facts." Such a double identity, as particular and universal, underpinned the increasing identification of the "true" with the privileged pole of the structures of knowledge (what would come to be known the value-neutral "sciences"), differentiated and dissociated from the "good" at the opposing pole (the value-laden "humanities").

By the sixteenth century, the inventions of value-neutral time (clocks) and space (grid-work maps) and the discovery of the Americas spurred humanists to propose alternate models of the heavens. Copernicus and Kepler placed the sun at the center of the universe. Giordano Bruno went further; he envisioned a virtually limitless universe filled with an infinite number of worlds. The disorientation that resulted from finding oneself adrift in an immeasurably vast cosmos only invited further efforts at reform, helping to make astronomy the most popular branch of natural philosophy at the time (R. Gascoigne 1992: 562).

Perhaps the greatest impact of humanism, however, was its role in the shattering of European religious and political unity in the course of the Reformation. Its emphasis on the centrality of the text ultimately undermined the church as the intermediary between God and humanity in favor of scripture itself and resulted in the coming to the fore of literalism and fundamentalism on both sides of the religious divide (Eisenstein 1983: 151–53). Furthermore, the attack mounted on the Roman Catholic Church, first by Luther and then by other Reformers, destroyed Christendom's sense of community and common destiny, thereby opening the spheres of politics, history, and jurisprudence to secularization (Voegelin 1998). The fact that some religious authorities now claimed a monopoly on the correct interpretation of the Bible while at the same time attempting to silence their competitors radically altered the intellectual map of Europe. Censorship and indexes of prohibited books were commonplace in both the Catholic and Protestant camps. The gradual gravitation of authors, printers, and secularly inclined natural philosophers to the Protestant centers in the north had more to do with the limited spread of the Reformers' authority than with the purportedly greater openness of Protestant theology (Eisenstein 1983: 170–73, 229–52).

The success of the humanist movements was in large measure driven by the conditions of economic, political, and technological expansion. In the commercial cities of northern Italy, the largest in Europe, ecclesiastical institutions were simply unable to address the demand for practical education, and as a result it was humanist learning that accounted for the relatively high literacy rates that prevailed in this region (Grendler 1989: 6–11, 403). Through the medium of the Reformers' propaganda campaigns, print vernaculars and lay literacy also spread north of the Alps (Edwards 1994). Ambitious Renaissance princes supported the spread of the new learning because association with the humanist virtuosi offered them an opportunity to proclaim "the power and grandeur of the court either symbolically or by measuring the extent of personal political authority and wealth" (Moran 1991: 171). Finally, the introduction of printing in the fifteenth century triggered an exponential increase in the number of available texts and allowed for continuous updating and correction of information.

The Search for Unity

In the wake of the religious wars that raged across Europe, the restoration of order in the domain of knowledge became the chief aim of most major intellectual movements (Shapin 1996: 119–65). The fundamentalisms, which had monopolized the political sphere and taken control of social movements, had become too divisive. Settling philosophical disputes by recourse to direct observation or to logical and mathematical arguments, at least in the world of nature, became common among partisans, both Jesuit (Dear 1995: 6) and Calvinist (Daston 1992). The authority of texts was being eclipsed by a reverence for the Book of Nature. The medieval cosmos, with its miracles and qualitative distinction between the terrestrial and celestial spheres, was giving way to the worlds of Kepler and Galileo, where nature acted with lawlike regularity, where phenomena such as magnetism could be found in either realm, and where mathematization was universally applicable.

Proponents of new mechanistic philosophies pictured nature as God's clockwork, fit to be studied in the same way one analyzes the workings of an artifice. This "maker's knowledge," including both mechanics and geometry, endowed the philosopher with a higher degree of certainty than knowledge possessed by observers or users because it "presuppose[d] a putative capacity for re-enactment by others" (Perez-Ramos 1988: 49). The Hermetic revolution spearheaded by Bruno and the Rosicrucians around the turn of the seventeenth century rallied these intellectual movements as heralds of a new Golden Age in which human beings would reclaim their birthright as operators endowed with nearly godlike powers. The Hermetic promotion of the pursuit of knowledge as a heroic quest resonated widely throughout the next century, the so-called Century of Genius. But politically, it proved no less destabilizing than religious fundamentalism. Henceforth, natural philosophy would be increasingly defined by its placement of strict limits on investigation and by its explicit exclusion of religious and philosophical doctrine from its method (Westfall 1977: 13–42).

Though often counterpoised to one another, both Francis Bacon and René Descartes became fellow travelers in the advancement of the mechanistic epistemological program (Perez-Ramos 1988; Gay 1964: 11–12). Bacon attacked both classical philosophy and Christianity as constraints on human potential, advocating instead the employment of "the Divine gift of reason to the use and benefit of mankind" and arguing that "the real and legitimate goal of the sciences, is the endowment of human life with new inventions and riches" (F. Bacon [1620] 1899: 23, 339). Furthermore, he advocated a strict separation between the divine and human branches of knowledge, asserting that the latter must adopt mechanistic principles and reject teleology. Suspicious of the imperative to quantify all knowledge, Bacon visualized natural philosophy as a cumulative, long-term endeavor based on direct observation and the maintaining of a catalog of doubts, errors, and irregularities (F. Bacon [1620] 1899: 22, 93). But despite its clear central role, science was to remain carefully secreted from the majority and subjected to the moderating influence of classical philosophy and orthodox Christianity (Weinberger 1985: 32).

Descartes, a geometer and systematizer, objected to the Aristotelian division between nature and artifice and the assumption that our senses are adequate perceptors. His own distinction of mental (*res cogitans*) and material (*res extensa*) objects did not stand in the way of his contention that the objects studied by mathematics and by natural philosophy were identical (Larmore 1980: 12–13). The clear and distinct perception of natural phenomena depended on a priori principles that made it possible "to know certainly and fully countless things—both about God and other intellectual matters, as well as about corporeal nature, which is the object of mathematization" (Descartes 1980: 89). The Cartesian method explained any effect as caused by direct physical contact through the medium of a particulate ether. It was more solipsistic and less collectivist than the method proposed by Bacon, thus exemplifying what might be described as the authoritarian ambitions of mathematicians to "absorb the cognitive territory of the natural philosophers" (Dear 1995: 8). Nevertheless, like Bacon, Descartes strove for restoration of order and recommended subjecting to moderation, law, and custom all things that could not be clearly and distinctly perceived (see Gaukroger 1980: v; Descartes 1968: 45).

Informal groupings dedicated to the advancement of the new natural philosophy sought royal support and by the 1660s succeeded in chartering new types of institutions, of which the Royal Society of London and the Académie des Sciences in Paris were the earliest examples (Purver 1967; McClellan 1985: 44–65). The academies offered opportunities to come into correspondence with a wide variety of other scholars and to verify the results of experiments, as well as (occasional) access to funding and publication in official periodicals, which became the most important impetus for the expansion of the scientific community since the printing revolution (R. Gascoigne 1992: 545). In contrast to their Renaissance predecessors, these scholars enjoyed a corporate status that allowed for perpetuation beyond the life of an individual benefactor (McClellan 1985: 3). As such, they could own land, operate a printing press, and in certain cases, enjoy exemption from taxes and censorship. Another key innovation was the limiting of academic business to matters of natural philosophy and the explicit prohibition against discussing contentious topics like politics and religion at official meetings (Lyons 1968: 41, 329–40; Hahn 1971: 11–12, 58–66). This was a clear attempt to institutionalize the separation of the production of authoritative knowledge from human values, even though it was largely the "value" accorded to gentlemanly conduct that underwrote any confidence in the findings of experimental natural philosophy (see Shapin & Schaffer 1985; Shapin 1994).

With the spread of the new institutions throughout most of Europe (and beyond) in the course of the eighteenth century, two distinct models emerged. In countries with strong parliamentary bodies (Great Britain, its North American colonies, the United Provinces of the Netherlands), natural philosophy was practiced by *societies*, whereas in the absolutist monarchies of the Continent the more elitist *academies* predominated (McClellan 1985). In the former, which remained largely independent of royal control and which opened admission to large numbers of amateurs to support their activities, it became a gentlemanly pursuit, part and parcel of cultivation, public-spiritedness, politeness, and patriotism (Golinski 1992; Sutton 1995; Widmalm 1992). The new ("Baconian") experimental sciences of chemistry, magnetism, and heat prospered in areas where experiment, fact collection, and utility appeared as the essence of civilization. On the other hand, the Continental academies, which generally excluded the lower orders and ennobled members to reinforce the traditional social hierarchy, displayed a more courtly ethic, stressed grand gestures and virtuoso performance, and specialized in the more traditional sciences, astronomy and mechanics, which remained accessible only to experts with extensive mathematical training (McClellan 1985: xxviii; Hahn 1971: 35; Biagioli 1992). Thus estranged from civic life, the savants tended to find fulfillment in purely intellectual endeavors or in the pursuit of social advancement (Keohane 1980). But in the leading academies, which were divided into specialized sections and in which election to membership was tightly controlled and the research agenda set to reflect a utilitarian *raison d'état*, these *honnêtes hommes* became state bureaucrats and the first professional "scientists" (Hahn 1971: 17, 22, 58–81).

Institutionalized natural philosophy tended to succeed best in countries with strong commercial classes and a centralized monarchy. As it offered a relatively free exchange of ideas in a space where traditional status hierarchies did not apply, the former found it appealing because it offered opportunities for social advancement and self-expression (Ben-David 1971: 75–87). Since it endeavored to build consensus on the basis of certain, nonpartisan knowledge, it also proved an attractive tool in states' efforts to transcend the fractiousness of the wars of religion and to affirm their own legitimacy (Shapin & Schaffer 1985; Hahn 1971: 9). In the writings of Hobbes, the German cameralists, and the physiocrats, it even became absolutist ideology. Furthermore, the alliance between monarchy and natural philosophy counteracted the internationalist and antihierarchical tendencies of the Republic of Letters (Widmalm 1992: 248–49; Hahn 1971: 44–45). Though technical applications of the new knowledge still remained distinctly secondary reasons for royal support, states (as well as private individuals) began to sponsor prize competitions at academies and societies for the purpose of solving practical problems (Bektas & Crosland 1992; Hahn 1971: 62, 120). Subsequently, what came to be known as science developed fastest in countries where the forces of Latin universalism and regional particularism were successfully countered. Thus, England and France, where science became institutionally and linguistically nationalized, outpaced the United Provinces, Sweden, Scotland, and Germany, where the universities persisted as the major centers of learning (Ben-David 1971: 111; Cook 1992) and where internationalist models still often continued to hold sway in the first half of the eighteenth century (Hertz 1962: ch. 6).

The increasing divergence of the trajectories of European and non-European structures of knowledge was favored by the contemporaneous failure of empire and the conquest of the Americas, which stimulated a continually high level of competition between medium-sized states. In fact, the number of knowledge workers engaged in what we now consider science began to increase exponentially after the introduction of printing in the 1450s, and the trend continued throughout the four subsequent centuries (R. Gascoigne 1992). Although even into the eighteenth century "scientific practice constituted a dependent variable, something that followed from other activities" (Lux 1991: 186), the structural hierarchy of facts and values, realized in the two-cultures antinomy, already constituted the long-term basis for deciding the authority and legitimacy of knowledge claims in the modern world-system.

Consolidating the Geoculture of the Modern World-System

By the middle of the eighteenth century, the empirical and political successes of natural philosophy evidenced its role in geocultural transformation. The institutionalization of academies in the previous century prepared fertile ground for the dissemination and testing of hypotheses and their subsequent transformation into paradigms accepted by the overwhelming majority of practicing natural philosophers.[1] The first such paradigm arose on the basis of Isaac Newton's work, which in terms of method brought together Baconian empiricism and Descartes's deductive mode. The model was especially successful in mechanics, in particular Newton's theory of universal gravitation (B. Cohen 1985: 161). When it received empirical confirmation, the idea that certain, indisputable knowledge, in the form of universal laws, was in principle obtainable began to have an impact on the sphere of human relations as well. The leaders of the Enlightenment movement called for "man's emergence from his self-incurred immaturity" into the Age of Reason (Kant [1784] 1970: 54) and for the reconfiguration of social institutions in line with Newtonian natural philosophy (Gay 1969: 126–66). When they applied the Baconian model of the incremental accumulation of useful knowledge to history, they arrived at the theory of progress—the idea of the constancy and desirability of social change for a better world (Bury 1932: 162). By century's end, this theory was deployed for the purpose of legitimating the new revolutionary republics, and it acquired geocultural force when it was tacitly adopted by their enemies as well (Wallerstein 1991a: 8–9). "Fear of change, up to that time nearly universal, was giving way to fear of stagnation" (Gay 1969: 3).

In the process traditionally denominated as the scientific revolution, Isaac Newton was in fact a transitional figure who is now regarded as the "last of the Renaissance magi" (Figala & Petzgold 1993; Webster 1982). Newton followed both Bacon and Descartes in stressing the limitations of natural philosophy. His own method involved balancing a variety of approaches, including mathematics, experiment, observation, alchemy, and religious history, against one another in order to ensure that his pursuit of the truth would be self-correcting (Teeter-Dobbs & Jacob 1995: 10). Newton rejected Descartes's theory of ether as potentially atheistic and denied that mechanistic hypotheses sufficed to explain the workings of complex chemical or organic structures. Believing that the "geometric is nothing but the perfectly mechanical" (Dear 1995: 213), he proposed instead that mechanistic philosophy be limited to the application of mathematics to determine the formulas of attractive (or repulsive) forces (such as gravity) that act on matter at a distance. He affirmed (contra Descartes) absolute time, space, mass, and force, whose ultimate reality could not be comprehended, for all were creations of God (Westfall 1977: 159).

However, in the 1730s, when the collective work of all the major European academies established that Earth was an oblate spheroid, as predicted by Newton's theory of gravitation, and not round, as postulated by Descartes, Newtonian mechanics were set apart from the rest of his method, and the universe was reconceptualized in purely mechanistic terms: "The term Newtonian was now applied to everything that dealt with a system of laws, with equilibrium" (Prigogine & Stengers 1984: 29). Newton himself was lionized as the Moses who had led all of Europe out of the desert of ignorance, although the mechanization of nature associated with his name had been greatly influenced by the cultural expectations fostered by the Continental rationalists (Gay 1969: 128–50). Stimulated by the creative tension initiated by Newton's refutation of much of the substantive work of Descartes, it was the more deterministic, French variant of natural philosophy that demonstrated the most dynamism in the second half of the eighteenth century (Brockliss 1992: 78–79; Ben-David 1971: 83–87). Henceforth, the essence of natural philosophy (and thereafter, science) would reside in the derivation of causal laws, the positing of which depended on the supposition of identity between a constantly changing lived-in reality and a static model of that reality (Meyerson 1930: ch. 6). This operation allowed one to apprehend the enduring relations that underlay the fleeting objects of our everyday experience and opened the way for rigorous quantification. By the end of the eighteenth century, Immanuel Kant popularized the notion that "in every special doctrine of nature only so much science proper can be found as there is mathematics in it" (Kant 1970: 6).

The triumph of Newtonianism had distinctly different outcomes in different countries. In England, the success of Newton's followers in integrating mathematics (but not experimental science) into the traditional university curriculum reinforced the compact between the post-1688 limited monarchy, the Anglican Church, and natural philosophy (Teeter-Dobbs & Jacob 1995: 64–71; J. Gascoigne 1989). But at the same time, the openness of the philosophic societies, civic institutions, Dissenting academies, and the universities in Scotland allowed technically inclined Newtonians to propagate the new knowledge among skilled artisans, who applied it to the industrial development of steam power (Cardwell 1963; Teeter-Dobbs & Jacob 1995: 71–78).

Elsewhere (but especially in France), admirers of England's stability and prosperity converted to the mechanics of *le grand Newton* and the sensationalist psychology of Locke and demanded their rigorous application for the purposes of reconstructing the social order (Gay 1964: 308–21; Gay 1969: 126–66). Attacking metaphysics, namely the Cartesian corpus, which had by then been re-Scholasticized by the Jesuits (D. Clarke 1989: 231–41), and religion, namely the Catholic Church, as the founts of idle speculation and hence as the promoters of ignorance, the philosophes[2] proposed to replace them with natural history, the secular history of humanity, and the "sciences of Man" (White 1970). These latter were constructed on an exclusively naturalistic or mechanistic basis, eschewing ethical or aesthetic commitments in favor of scientific objectivity, cultural relativism, social construction, and the pleasure/pain calculus (Voegelin 1975: 35–52; Gay 1969: chs. 4, 7). The popularization of their agenda in the *Encyclopédie* made the Enlightenment of the philosophes common property of the European educated classes (Thrift 1996: 112–13), while its radicalization in various republican, pantheist, and Masonic circles transformed it into a potent advocate for the overthrow of the ancien régime (Jacob 1981).

The generalized success of Newtonian mechanics lent prestige to deterministic worldviews, deepened the two-cultures split, and furthered the establishment of the pursuit of value-free knowledge as the geoculture of the modern world-system. By century's end, Pierre-Simon Laplace's replacement of God with a universal intelligence capable of calculating every cause of every effect opened the way for the complete determination of natural systems. No longer bound by a metaphysical necessity to remain content with mere probabilities, the human scientist became the final arbiter of truth (Hahn 1967).

Conversely, the proliferation of mechanistic models in chemistry and the life sciences, until then largely taxonomic or dominated by vitalists and naturalists, elicited a backlash (Kiernan 1973: chs. 5–7), in no small part because these fields retained independent institutional bases at university medical faculties and botanical gardens. Envisioning nature as imbued with a creative principle rather than passive, law-bound, and fit only to supply standards for aesthetics, morality, feminine behavior, and child rearing,[3] the vitalists rejected the identification of chemical structures and living organisms with particular arrangements of atoms. The defense of vitalism within the organic sciences played a key role in stimulating the romantic movement (Kiernan 1973: 45–46), which would challenge the idea that the study of humanity in particular could proceed in a purely mechanistic framework.

In due course, however, the best efforts to maintain even a loose unity between knowledge of the natural world and modes of understanding the human world foundered, and even Kant's conciliatory efforts ended up postulating two kinds of reason—pure and practical—thus contributing to the rigidity of the growing division between materialism and idealism (Barbour 1997: 46; Prigogine & Stengers 1984: 89).

The inculcation of a culture of quantification, precision, and discipline helped the developing industrial structures (especially in Great Britain) to control an expanding labor force in an increasingly sophisticated technological environment (Thompson 1991: ch. 6; P. Cohen 1984). The economic imperative to standardize weights and measures produced what has been called a "revolution in measurement"; reform proved particularly thoroughgoing in France, where the revolutionary government adopted the metric system, long sought by academicians who demanded a precise and uniform system that took its standards from nature (Zupko 1990). When the French universities and academies were reorganized and centralized as the Institut de France in 1795, natural scientists became, for the first time, public servants. The application of scientific knowledge to the business of state was promoted by the founders of new republics, not only in France (most radically in the form of Condorcet's social mathematics) but also in the United States (Crosland 1967: 90; Gay 1969: 560).

The claim of managing the state in accordance with scientific principles brought international prestige, prompting Napoleon to exalt scientific knowledge above all others and to restrict membership in the elite first class of the institute to natural scientists alone (Crosland 1967: ch. 1). The new *grandes écoles* (most notably the Ecole Polytechnique) created institutional opportunities for the reproduction of scientific knowledge. Although during the Bourbon Restoration many of these changes were temporarily reversed, Napoleonic France managed to export Enlightenment ideology (Saldaña 1992), institutional reforms (C. McClelland 1980: 101–4; Pavlova & Mikulinskii 1990: 45–62), and mathematical physics (Crosland & Smith 1978; Dauben 1981) to the rest of Europe and the Americas.

Having captured the leading position in world science (Ben-David 1971: 88–107), France perpetuated a system in which a semiperiphery (southern and eastern Europe) became dependent on it for the training of personnel, while a periphery (Latin America and the colonial world) assumed the role of a natural laboratory and a supplier of specimens (Basalla 1968; Polanco 1992). To escape scientific peripheralization, Russia pursued the strategy of importing a complete Western educational and scientific infrastructure (McClellan 1985: 76–82). All peoples who failed to demonstrate a given level of scientific development were becoming devalued in European eyes (Adas 1989: ch. 2).

Resistance and Specialization

The romantic movement, arising largely out of opposition to the Enlightenment, likewise influenced the development of scientific culture. Originating primarily in Germany (where Newtonianism never really took root), it was spearheaded by the *Naturphilosophen,* who insisted that true science should study the world as a whole instead of specializing in its mechanistic aspects. The reassertion of speculative philosophy directed the romantic conception of science to be sensitive to "the need of a global, systematic and solid picture of knowledge of nature . . . and at the same time [to] the importance of the non-preconceived observation of phenomena" (Poggi 1994: xiv). Believing that the Newtonians mistook their mechanistic abstractions for reality itself, critics like Johann Wolfgang von Goethe seemed to recommend a more Baconian approach focused on "increasing . . . knowledge of the living phenomenon in all its given contingency" (Stephenson 1995: 10) over the abstract method centered around the testing of theories. The romantic stress on empirical observation and holistic contemplation challenged the Newtonian attempt to subordinate the universe to a single, reductive system of laws, thereby encouraging the spread of new antimechanistic sciences and compartmentalization in the structures of knowledge. In contrast, the synthesis of Baconian and Cartesian strands evolved into a universal scientific method, which (unlike the practice of deducing an object's properties from previously established principles) "accepts an unknown, yet unexplored object or problem as given, and reduce[s] it by help of all the relationships and relations this quantity must satisfy, to known quantities" (Jahnke & Otte 1981: 80). As the individual disciplines became institutionalized and professionalized, this method unified the natural sciences as a distinctly demarcated domain.

The sciences of the period developed in the crucible of the romantic reaction and continued to exhibit traces of the confrontation with Newtonianism long after the flames of romanticism subsided. In chemistry, the anti-Newtonian imperative largely accounted for its declaration of independence from mathematical physics and its subsequent prominence as the most significant science of the first half of the nineteenth century. In the 1780s, Antoine Lavoisier, although he dismissed phlogiston as a substance (because it had no measurable properties) and introduced sound gravimetric and experimental practices, nevertheless sought to redeem the efforts of eighteenth-century vitalist chemists by ensuring that chemistry would remain a separate science devoted to the study of qualitative *affinities* of specific elements, rather than reducing them to the generic and purely quantitative Newtonian forces (Gough 1988; Nye 1992: 208–10). Subsequently, under the dominance of Justus Liebig's school of organic chemistry, "during the mid-decades of the nineteenth century, chemistry appeared to be developing into an increasingly non-mathematical science" that focused on the practical work of classifying and synthesizing new elements and compounds rather than on developing a full-fledged theoretical framework of the kind pursued by mathematical physics (Nye 1992: 211; Nye 1996: xv).

In organic sciences like physiology, the influence of romantic thought was even more pronounced. In the early stages, physiology especially was heavily influenced by the speculative theories of Lorenz Oken, who rejected experimentation in favor of the study of archetypal forms and the vital principle imbued in all organisms. Even after empiricists had routed the *Naturphilosophen* and set about the study of cell chemistry, residual vitalism still had an impact on the work of pioneers like Claude Bernard, who maintained that "in every living germ is a creative idea which develops and exhibits itself through organization" and therefore "the domain of life . . . belongs neither to chemistry nor to physics" (quoted in Gasking 1970: 163).[4] Meanwhile in mathematics, the rethinking of Euclidian geometry and a renewed emphasis on rigorous proofs signaled an introspective turn and growing differentiation of pure mathematics from the Newtonian physicomathematical corpus (Grabiner 1981; Bottazzini 1994).

The romantic variant of science, as well as the distinct disciplines, was institutionalized in the resurgent German universities, which, in the absence of a large middle class and a sizable network of academies, themselves became the vanguard of secularization in Germany (Ben-David 1977: 14; Ben-David 1971: 110–11). Characteristically, this movement was initiated by arts and philosophy professors, who demanded equality with the professional faculties and invoked *Lehrfreiheit*—the right to pursue their own research interests and the freedom to construct their own graduate seminars. In response to these calls for reform and to the French challenge, Wilhelm von Humboldt founded the new University of Berlin in 1810 as a place where scholars could devote themselves fully to a scientific calling and the pursuit of pure knowledge. But the German conception of science—*Wissenschaft*—as formulated in the eighteenth century denoted any theoretically informed systematic study based on a critical treatment of sources and directed toward the augmentation of knowledge in one specific area. Therefore it did not imply, at least at first, any categorical distinction between the faculties of philology, history, or chemistry (C. McClelland 1980: 34–43). Only in the 1820s, when

German chemists returned from apprenticeships in Paris and Sweden, was laboratory work instituted in chemistry seminars, and chemical research redefined as the practical work of synthesizing new compounds (Rocke 1993: 13–34; Nye 1996: 9–11). In 1835, Liebig, who was to become the leading German chemist, succeeded in having his laboratory at Giessen officially recognized (and funded) as part of the university. Subsequently, and unlike in France, the researchers themselves propagated individual disciplines through the teaching of both theoretical foundations and laboratory techniques in a single institutional setting.

The synthesis of the neohumanist and empiricist ideals within the German establishment shaped the development of the scientific profession. Its research ethic centered on individual development and the heroic investigation of a vast and dynamic world—noble ideals to be pursued without heed of practical applications or social pressures (C. McClelland 1980: 118–24, 172). The researchers themselves were professors, who awarded doctorates of philosophy and were paid as pedagogues whose purpose in the laboratory "was not to show students how to boil soap or to compound drugs but rather to educate the mind and teach the student how to think" (Rocke 1993: 30).

But the presence of the laboratory on university premises—a visible intrusion of industry into the domain of knowledge (Nye 1996: xvi)—symbolized the development of the *differentia specifica* of professional science. First, laboratory science was committed to the study of nature in a controlled environment (James 1989: 3). Second, as reports of laboratory findings, scientific publications displayed a dry, technical style, and because the findings were products of teamwork, they often bore multiple authorship. The contrast with the quite literary writing of academic science in the previous century, as well as with contemporary natural history (which remained aloof from laboratory work and continued to be dominated by amateurs), reflected the professionalization that had taken place in the laboratory sciences—first chemistry and, later, physiology and physics as well (Manten 1980: 13).

Already by the 1840s, the German university model proved more successful than the French system, and in the second half of the century it was widely emulated. Initiated by Humboldt as part of a campaign of national rebirth in the face of French imperialism, university reform was supported by the middle strata because it allowed degree holders to establish monopolistic control on access to professional positions (Ben-David 1977: 19–20). As such, it proved attractive to other states (such as Russia and Japan) whose pursuit of upward mobility in the world-system had to be balanced against the necessity of fostering national integration and simultaneously maintaining strong central control and rigid class distinctions (J. McClelland 1983; Watanabe 1976). Though state-run, the German university system reaped the benefits of the lack of political centralization (until 1871), because each small state endeavored to have its own university and thereby augmented the size and competitiveness of the overall German scientific establishment.

By comparison with Germany, other established scientific powers began to exhibit a lack of dynamism. The main competitor, French science, suffered from excessive centralization. Owing to the separation of teaching and research functions, it proved increasingly incapable of competing with well-funded and well-staffed German laboratories (Ben-David 1977: 22–25). Its image abroad was further complicated by its association with dangerous revolutionary ideas. In Great Britain, where throughout the nineteenth century continued dominance by amateurs and a laissez-faire attitude on the part of the state raised concerns about falling behind Continental competitors, aspects of the German model began to be replicated as early as the 1830s. The new "redbrick" universities and the British Association for the Advancement of Science (BAAS) adopted the German commitment to specialized research and soon displaced Oxbridge and the Royal Society as centers of scientific activity (Cardwell 1957: 33–47). Shortly after the foundation of the BAAS in 1831, William Whewell coined the neologism "scientist" to describe those who attended its meetings (Nye 1996: 13). Thereby, the very word science assumed a new and clearly articulated meaning (utterly different from that connoted by *scientia*), and the geoculture of the modern world-system ceased to be a domain without a name.

Industry and the Cult of Science

In the middle decades of the nineteenth century, the reversals of the romantic reaction paved the way for a reaffirmation of mechanistic and instrumentalist views of science. Although the scientists' prerogative to conduct research free of all interference remained inviolable, they now acquired a responsibility to direct it to socially useful ends. Even in Germany, educators began to stress realism and empiricism and to link the Humboldtian research ethic's overemphasis on idealism and subjectivity with the fantasies of 1848 (Brain 1991: 369–72). More fundamentally, both Auguste Comte's heralding of the positivist age[5] (Comte 1974) and Marx's claim that the only science deserving of the name was that which "transformed human life *practically* through the medium of industry" (Marx [1844] 1964: 142; italics in the original) targeted the fragile resurgence of philosophy, a product of the romantic period. These revolutionary pronouncements soon found resonance among some scientists who, like Herman Helmholtz, proposed to reestablish philosophy as a theory of scientific knowledge based on the physiology of the senses (Moulines 1981). But having identified the real solely with material existence, the proponents of subjugating philosophy to science could not rest there. By the 1870s, the conception that science was in the process of solving all the riddles of the universe and had thereby "occupied every corner of space and left no room for God" (Chadwick 1975: 178) had evolved into the dominant view of the age.

Particularly indicative of the positivist reorientation of science is the history of the formation of modern physics. Though it has been asserted that nineteenth-century industry utilized primarily eighteenth-century science and often held back the application of more revolutionary theories (J. Bernal 1953: 149), steam engine technology clearly stimulated the rise of the sciences of energy (Cardwell 1963: 89). At first, under the impact of romanticism, vitalist notions like *vis viva* ("living force") invaded the study of molecules, magnetism, heat, and electricity, while the proliferation of different explanations for these various phenomena threatened the unitary quality of the system of natural laws that mechanistic natural philosophers had affirmed in the eighteenth century (Nye 1996: 88–94). Newtonianism suffered further setbacks as a result of the suggestion by Augustin-Jean Fresnel and Thomas Young that light, rather than being composed of particles, was actually a vibration in a fluidlike ether. In addition, Michael Faraday's discovery of curved electromagnetic fields repudiated the classical theorization of forces-at-a-distance as acting in every instance along a straight line across the shortest distance (Nye 1996: 62). Finally, in thermodynamics, the impossibility of identifying heat as a substance and the absence of a countereffect to the transformation of kinetic to potential energy (as stipulated by Newton's Third Law) indicated that, unlike mechanistic systems, thermodynamic systems were irreversible and could not be defined in terms of matter and forces.[6] But since the law of energy conservation stipulated that different forms of energy were convertible into one another, it also provided the axis around which the sciences of electricity, magnetism, optics, heat, and mechanics ultimately coalesced by 1870 to form the discipline of physics (Knight 1986: 159–62). Yet the need to express the quantity of conserved energy as useful work based on the absolute units of time, space, and mass in order to make the new sciences of energy commercially viable essentially amounted to requiring their restatement in mechanistic terms, and thus in their reconfiguration as the rightful heirs of Newtonianism (C. Smith 1998: 192–96, 268). The traditional marking of James Clerk Maxwell's establishment of the Cavendish laboratory at Cambridge in 1870 as the birth of experimental physics represents the victory of the utilitarian British approach of Sir Joseph John Thomson and Peter Guthrie Tait over that of the more theoretical Helmholtz and Rudolf Clausius in Germany (C. Smith 1998: 197, 285–86). Having successfully postulated new laws in, and extended mathematization into, the empirical fields heretofore contested by chemistry, physicists would now claim theoretical primacy within the sciences (Nye 1996: xv; Knight 1986: 174).

The intrusion of utilitarian considerations on the idyllic terrain of *Wissenschaft* broke the philosophically sanctioned moral community of researchers and ended the universities' monopoly on science. In its stead came diversification, both on the level of disciplines[7] and on the level of professional institutions (Jarausch 1983: 18–22). Since the pursuit of scholarly research agendas increasingly demanded more funding and sharing of expertise than universities could provide (much less the support networks of amateurs), scientists increasingly turned their attention to institutional organization and to the demonstration of their work's practical potential. No longer distinct simply by virtue of falling under the domain of different university chairs, after 1850, disciplines developed differentiated structures and identities via authoritative textbooks and maintained them through trade journals, professional societies, and specialized training (Lenoir 1993: 70–102; W. Paul 1980: 138; Nye 1996: 21–22). The proliferation of new disciplines (for instance, experimental physiology or pharmaceutical chemistry) specifically geared to the development of scientific laws with practical uses marked the birth of a division between pure and applied science, while disciplines concerned with applying these new laws in industrial production fell under the rubric of "engineering."

These three categories of fields became separate because they developed separate institutional bases. Pure sciences (primarily but not exclusively in Germany), while they remained within the universities, moved as well into private and government-funded research institutes (which became the new centers of scientific life) and gradually eroded the corporate relationship between the state and the university (C. McClelland 1980: 284–87). The applied sciences and engineering made some inroads in the universities, but the latter's resistance to "crass materialism" often forced these fields into *technische Hochschulen* and private establishments like the Pasteur Institutes (Lundgreen 1983: 160; Cardwell 1957: 124–34; Ben-David 1971: 104). In addition, the 1860s saw the rise of the industrial laboratory in German dyeing and pharmaceutical concerns, a trend subsequently manifested in other technologies and other countries. Faced with expanded enrollments and great institutional growth in the practical sciences, traditional university elites condemned scientific education as "a deal of trash."[8]

By the 1870s, the renewed confidence of the physical sciences and their by now undeniable impact on industry and society stimulated widespread imitation of physical models, a movement that in extreme forms began to resemble a secular religion. This was nowhere more evident than in the institutionalization of the disciplines of the social sciences. For instance, the founders of neoclassical economics appropriated (often uncritically and ineptly) the metaphors and mathematical models of energy physics (constructing utility as a field of potential energy) in an effort to reconstitute economics in the image of physics, as well as to intimidate rival social theorists—competitors for scarce funding—into submission (Mirowski 1989: ch. 4). Additionally, the prominence of the conservation of energy in the physical sciences led to the reconceptualization of the whole universe as a thermodynamic system, which, because it was in the process of heat transfer, was itself directional (and by implication, progressive) (C. Smith 1998: ch. 6). In the 1860s, this doctrine, defended primarily on religious grounds, was joined by the avowedly a-religious creed espoused by the proponents of Darwinian evolution. Whereas the Scottish physicists identified progress with the prudent and efficient use of the limited amount of God-given energy available to humanity, certain Darwinists maintained that natural selection was itself progressive through its institution of competition among species and individuals (Greene 1981: 7).[9] The evolutionist position gained increasing attention and notoriety in part because natural history remained accessible to the layperson (Merrill 1989) but also because it obviated the need for divine action in the creation of life, thus completing the banishment of God from science, which was made explicit from the time of Laplace (Barbour 1997: 73–74). In the wake of the condemnation of liberalism by Pope Pius IX in 1864, the opposition between science and religion was sensationalized, leading all political progressives to rally to the cause of anticlericalism and creating a cult of scientific materialism (Chadwick 1975: 111–12, 172–77).

As part of a new secular religion, science and technology were now enshrined as the defining features of European civilization and as justification for the conquest of "backward" lands (Adas 1989: ch. 4). All-embracing schemes to assign every people a ranking on the ladder of progress (with the Victorian gentleman naturally at the top) proliferated, while the efforts to find similarities between European women and non-Europeans (notably by measuring their skulls) reflected the imperative to establish a firm scientific basis for actual racial and gender inequality (Stepan 1996; Adas 1989: 292–318). At the same time, "[t]he gentle practice of the investigator"—science itself—was characterized as a civilized, manly pursuit in fundamental opposition to "the sickly dreams of hysterical women and half-starved men" (W. K. Clifford, quoted in Knight 1986: 193). With non-Western and female-dominated practices declared superstitions, Western medicine moved to extend the realm of science to the healing arts (de Zoysa & Palitharatna 1992; Ehrenreich & English 1973).

The rapid spread of science and scientism was enabled in great part by the massive expansion of industry in the second half of the nineteenth century. Technological breakthroughs (showcased with great fanfare beginning with the Great Exhibition of 1851) testified to science's power to help humanity transcend its natural limitations. Although for most of the nineteenth century industry depended on talented amateurs to supply it with new inventions and techniques, after 1860 large Continental enterprises increasingly sought researchers, engineers, and managers with scientific training (Nye 1996: 2–3; Cardwell 1957: 85, 181).

With the awarding of the first Nobel Prizes in 1901, the industrialists' valuation of science as the fount of useful inventions and discoveries won legitimacy among the public at large (Nye 1996: 27).

Equally important to the growth of scientific culture were the increasing corporatization and democratization of the state. Since science was seen more and more as a public service, the idea of state funding for universal education and the establishment of standards gained headway even in laissez-faire Great Britain (Cardwell 1957: 187). After widespread attribution of Prussia's military victory over France in 1871 to its superior educational system, state support for the sciences began to reflect strategic, not merely cultural, considerations (Knight 1986: 140, 209). On the one hand, state support for research in meteorology, astronomy, and seismology in the colonies underlined the role of the exact sciences as the bearers of civilization and turned the periphery into an arena of cultural rivalry among the Great Powers (Pyenson 1985; Pyenson 1993); on the other hand, journalists, political organizers, social reformers, and the new systems of mass education after 1880 promoted the spread of secularism and scientific knowledge, which they linked with social empowerment, to the working classes (Chadwick 1975; 12, 88–106; Reisner 1925). However, beyond Europe, both colonial and indigenous elites feared the corrosive effects of Western education and continued to restrict the expansion of scientific learning and institutions (Ihsanoglu 1992; Jami 1992; Adas 1989: 322–26).

The Hegemony of Science

Toward the end of the nineteenth century, scientists were increasingly reaching the conclusion that Maxwell's mechanistic synthesis in physics rested on shaky foundations. Despite its claims to explain all of physical reality by referring to a single set of laws, the synthesis in fact reinforced dualism by its differential treatment of matter and energy (Smith 1998: 300) as well as of corpuscles and electromagnetic fields (Einstein 1968: 90). In 1887, the Michelson-Morley experiment failed to find an empirical confirmation for the existence of the ether on which mechanistic explanations of electromagnetism ultimately depended, while the subsequent discoveries of radioactive waves, discontinuous energy emissions, and electronic mass further undermined monistic theories, stimulating an intense reexamination of the fundamental assumptions that lay at the heart of Newtonian mechanics.

Ernst Mach and Henri Poincaré challenged the notion that absolute standards of measurement existed, and they rejected the Laplacian universal mind on which absolute time and space depended as metaphysical and not in accordance with *observed* facts (Mach 1968; Dantzig 1968: 135–36). Shortly thereafter, the physical relativity of uniform motion and the variance of mass according to velocity would be demonstrated by Albert Einstein's special relativity and "[a]ccording to the general theory of relativity, the geometrical properties of space are not independent, but . . . are determined by matter" (Einstein 1961: 113). Moreover, Einstein also extended relativity to the conservation of energy. Matter had the propensity to perform work under local conditions, but no universal frame of reference from which conservation could be observed existed. In contrast to the claims of energeticists in the nineteenth century, energy now presupposed matter, not vice versa (Sachs 1988: 216).

But perhaps the most damaging blow to the traditional model of an absolutely predictable universe on the basis of objectively given space, time, mass, and forces was delivered by the new science of quantum mechanics as represented by the Copenhagen interpretation of Niels Bohr and Werner Heisenberg. In what became known as the Heisenberg uncertainty principle, they affirmed that the very act of locating electrons distorts either their position or their velocity. Since our choice about whether to treat electrons as waves or particles seemed to affect the very phenomena studied, the Copenhagen physicists proposed abandoning claims of certain knowledge and objective causality in favor of a radically empiricist positivism (Nye 1996: 177–78; Sachs 1988: 119).

In the aftermath of the First World War, the widespread disillusion with scientific civilization contributed to a revolt against determinism (Feuer 1982: 158; Adas 1989: 379–80). Nonetheless, after the tumultuous first decade of the interwar period, when the Copenhagen interpretation and a rival interpretation of quantum mechanics championed by Einstein contested for primacy, order in the physical sciences was gradually restored and claims of scientific determinism reestablished. Throughout the 1920s, Einstein argued that divisibility into countable parts (or quanta) merely served as an approximate description of matter, which was continous and endowed with objective existence. Bohr and Heisenberg, conversely, eschewed materialism and the pursuit of conceptual clarity in favor of the simplest possible mathematical approximations of physical relationships and upheld quantization and instantaneous (noncausal) action on the micro level (Nye 1996: 177–78; Sachs 1988: 3, 146–48, 182; Feuer 1982: 175).

The majority of physicists ultimately decided the issue in favor of the latter, in part because Bohr's model of the atom enabled more precise calculations for the values of atomic radii and electronic orbits (Nye 1996: 169) and subsequently proved useful in predicting nuclear structure, thereby helping to create the applied field of nuclear physics in the 1930s (Kevles 1977: 222–35). Nevertheless, the victory of the Copenhagen interpretation of quantum theory did not result in the displacement of alternative paradigms (most importantly, of relativity theory) in the physical sciences, and certainly not in the enthronement of radical empiricism as the dominant scientific philosophy. Bohr's own denial that reality required a meaning militated against the success of his theoretical beliefs as a dominant doctrine (Feuer 1982: 110, 140–41) and Einstein's tremendous authority ensured that the more determinist views he supported remained prominent. Indeterminacy, he argued, lies in our insufficient knowledge of the object, not in the object itself. The knowledge of physical laws remained absolute and, like the speed of light, independent of any observer (Ortega y Gasset 1968: 149; Nye 1996: 178).

Finally, in the 1930s and 1940s, a reaction to relativism took hold, partly in response to the global crisis, partly in virtue of the aging of the revolutionary generation of scientists, and partly because of the return to stability in the physical sciences, fostered by their practical successes and by the absorption of the new paradigms by academic structures (Feuer 1982: 78–84, 354–61). As the leading spokesman for the philosophy of science in the postwar period, Karl Popper agreed with Einstein in rejecting Heisenberg's and Bohr's antideterminism as itself deterministic in the sense that it could not be falsified (Popper 1959: 248–49). Although he strongly rejected the scientisms that dominated the late nineteenth century and considered the question of whether the universe is in fact subject to deterministic laws in principle unanswerable, Popper nevertheless reaffirmed the striving for such laws as the task of science (Popper 1959: 246), thus restoring to science alone the right to make authoritative statements on the universe.

Because physics began once again to appear orderly and successful, its ability to project its dominance over other branches of science and laypersons was reinstated, although the full implications of the changes that had taken place in the apprehension of the universe were often lost on them. As a result, other disciplines strove to establish greater conformity with physics, and, especially in the case of the social sciences, their borrowings were often uncritical or cosmetic and were cast in a decidedly deterministic light. In chemistry, the calculation of the energy of molecular bonds precipitated the proposition of the electron-bond theory of chemical affinity, thereby stimulating efforts to reconcile chemistry and physics and to reunify physical and organic chemistry (Nye 1992: 214–15; Nye 1996: 178–82). The latter in turn influenced the development of molecular biology, a field equally divided between physicists and chemists until the discovery of DNA in 1953 (Morange 1998: 12–20, 68–78). Because the paradigm of molecular biology had the field to itself (unlike the situation in physics), the physical sciences seemed to impart to molecular biology "a belief in the absolute adequacy, not simply of materialism, but of a particular kind of (linear, causal) mechanism; belief in the incontrovertible value of simplicity; belief in the unitary character of truth; and finally, belief in the simultaneous equations between power and knowledge and between virtue and power" (Keller 1993: 56–57). In economics, which in the wake of the Great Depression also became heavily inundated with out-of-work physicists, the borrowings from physics were

limited to mathematical formalisms, which were supposed to reaffirm
the scientific status of economics and to conceal its origins on the basis
of the nineteenth-century energetics metaphor (Mirowski 1989: ch. 7).

The conceptual transformations beginning at the end of the nineteenth century that characterized developments in the twentieth century
coincided with a structural reorganization of institutional science (B.
Cohen 1985: 92–93). Much of the practical and theoretical research that
contributed to the rise of quantum mechanics, relativity, and physical
chemistry took place at the new research institutes established prior to
the First World War. The coordination between scientists working at
these institutes and the new international commissions in charge of stan-
dardizing the products of research (e.g., the International Committee on
Atomic Weights, founded in 1897) fostered the growth of an international scientific network that exhibited increasing independence of the national infrastructures constructed in earlier periods (Crawford 1992:
38–46).

The connivance of scientists with their respective governments during the war halted the trend during the 1920s (Nye 1996: 191–220), a
development further aided by the fact that the construction of particle
accelerators, so necessary to further progress in nuclear chemistry and
physics, could only be undertaken with state financing (Cardwell 1957:
172). Subsequently, the internationalization of science proceeded under
the auspices of the United States, at first because no agency could com-
pete with the increasingly international reach of U.S. corporate foundations and then because U.S. universities became such a magnet for
European scientists fleeing political persecution, anti-Semitism, and war
in the 1930s and 1940s (Geiger 1986: 162–63, 240–45). By the 1960s, com-
mentators like C. P. Snow were estimating that the U.S. share in world
pure science was 80 percent or higher (Kerr 1994: 69).

Capitalizing on their social, geographic, and financial advantages,
research universities in the United States became the primary site for the
institutional development of scientific culture following the Second World
War. Though initially patterned on the German universities in the 1880s,
the large size of the United States, its rapidly growing population, and
its relative lack of social barriers to higher education (excepting against
women and racial minorities) prompted U.S. universities to focus on
public service and to remain open to innovation and applied research
(Ben-David 1971: 141–66; Herbst 1983). Owing to the large sizes of student bodies, university faculties in particular fields began to include
multiple professors who organized into departments, which, because
they relegated the definition of disciplinary boundaries to national professional organizations and publications, promoted disciplinary conti-
nuity to a far greater degree than did the system of individual chairs in
Germany (Geiger 1986: 36–38). Furthermore, this concentration of money and talent completed the long transition to the era of so-called big

science. The universities' embracing of practical fields (notably engineering, agriculture, mining, and medicine) attracted the attention of industry, and by 1920 corporate endowments outdistanced tuition as a source of revenue (Geiger 1986: 41–42). Private foundations shaped research methodology and, as is illustrated by the case of the Rockefeller Foundation's role in the history of molecular biology, often contributed to the birth of new disciplines (Morange 1998: 81). In addition, the foundation of the National Research Council in 1918 as a coordinating organ between industry, science, and the federal government became a watershed in the transformation of science from the domain of the isolated genius to a horizontally and vertically integrated "army" of organized and cooperating researchers (Noble 1977: 158–59).

The beginning of serious government investment in nuclear physics in the context of the Second World War (and especially the Manhattan Project) marked the definitive birth of big science. With the onset of the Cold War, wartime research enterprises were transformed into permanent national laboratories (Kevles 1977: 367), while yearly government allocation of research and development funds in all fields to university-run laboratories grew from $310 million in 1940 to over $4.5 billion in 1968 (40 percent of which was earmarked for defense). The fact that the social sciences received only 3 percent of these funds and the humanities less than 1 percent testifies to just how "big" (natural) science had become (Kerr 1994: 40–41, 142). Between 1945 and 1968, the only notable competition for U.S. big science came from the USSR, which, while it succeeded in matching U.S. accomplishments in nuclear science and space exploration, proved less able to encourage innovative research in most other areas owing to the excessive centralization of its scientific establishment, the separation of teaching and research functions between universities and institutes, and poor contacts with Western scholars (Ashby 1947: 17–40; Medvedev 1978: 60–88).

The emergence of big science was telling proof of the leading role of scientific culture within the structures of knowledge in the modern world-system. The restoration of world order under the aegis of U.S. hegemony, along with the practical successes of the 1930s–1950s, restored the confidence in science that had been forfeited after the First World War (Adas 1989: 402–19). The unbounded exuberance with which science had earlier confronted religion was now replaced by more cautious attitudes, such as those expressed in Popper's claim that progress occurred as a result of skepticism and the falsifiability of individual theories (Nye 1996: 228–29). At the same time, the branding of unconfirmed laws and unfalsifiable theories as pseudoscientific only strengthened the authority of physics and recommended the adoption of its methodology in all other fields (Popper 1961). Similarly, the discipline of the laboratory became applicable to any other workplace in the form of "scientific management" (Bendix 1956).

As a globally hegemonic culture, science had no absolute, unitary form but rather remained an expression of the two-cultures structure while exhibiting local variations in semiperipheral and peripheral zones. The intertwining of science with the philosophy of dialectical materialism in the Communist bloc and the claim of dialectical materialism to absolute superiority over any other form of knowledge or system of thought displayed much greater similarity to the nineteenth-century model than its counterpart in the twentieth-century West (Ashby 1947: 97–126). In areas emerging from under colonial rule, science was promoted as the guarantor of true liberation and economic development (Eban 1961) and counterpoised to the fatalistic and passive characteristics attributed to non-Western knowledge cultures (Inkeles 1969).

Yet, in whatever variety, as of circa 1970, scientific culture defined authoritative knowledge—for very many it defined the only form of knowledge—everywhere in the world.

Notes

1. Kuhn (1962, 10, 37, 162–67) defines paradigms as those institutionalized practices and theories that are dedicated to the solutions of particular problems and that have attracted an enduring group of adherents and developed into traditions. In contrast to the social sciences and the humanities, the sciences generally allow only one paradigm in any given field.

2. Gay characterizes the philosophes as a "loose, informal, wholly unorganized coalition of cultural critics, religious skeptics, and political reformers from Edinburgh to Naples, Paris to Berlin, Boston to Philadelphia [which] united on a vastly ambitious program . . . of secularism, humanity, cosmopolitanism, and freedom . . . in its many forms" (1964: 3). In this, they differed markedly from the establishment proponents of Newtonianism in England, where the Enlightenment is usually seen as characteristically absent because of the relative lack of cognitive dissonance among elites (J. Gascoigne 1989).

3. The feminization of nature in the writings of the founders of mechanistic philosophy has been underlined by Lloyd (1996) and Keller and Grontkowski (1996), among others.

4. The pursuit of an independent and nomothetic organic science, as well as the very name biology, has also been linked to the professionalization of modern medicine around the turn of the nineteenth century (see Salomon-Bayet 1981).

5. For Comte, positivism denoted a rejection of the search for final causes, acceptance of the relativity of all knowledge, and the derivation of laws on the basis of an experimental, observational, and comparative approach (Comte 1974). Fashioned under the influence of Laplace and designed as "a program of unification of scientific explanation based on mathematical principles," it succeeded in establishing a following in the physical sciences (Nye 1996: 26).

6. *Energy,* defined by Joule, Rankine, and Thomson in the 1840s and 1850s as the capacity to perform work or to produce motion in bodies, differed from Newtonian *force* in that it underwent a *qualitative* transformation in passing from potential to kinetic states. The conserved energy could be measured, but its productive capacity was lost irrevocably (see Prigogine & Stengers 1984: 108; Smith 1998: 110).

7. In contrast to *fields*, which are centered around research programs oriented to solving particular problems, *disciplines* have been defined as "fundamentally institutional in terms of orientation . . . concerned more with establishing service roles, facilitating links to other disciplines, and enabling transmission of the techniques and conceptual tools of the scientific field to . . . user groups from neighboring disciplines and to persons training for particular types of careers" (Lenoir 1993: 79).

8. Cardinal Newman, quoted in Kerr (1994: 2). Though the opposition between traditional educational establishments and the new practical sciences took on a particularly sharp edge in England, where the Oxbridge establishment remained aloof from German innovations and where a professional rank and file remained absent until the end of the century (see Cardwell 1957: 49, 111–19), reactions of a similar nature can be found in the pronouncements by German university chemists, pronouncements that often took the form of anti-Semitism (Rocke 1993: 350–53). In Russia, on the other hand, tight state control and the small size of the scientific establishment ensured that profound cultural divisions would exist between czarist officialdom and the revolutionary intelligentsia (Pavlova & Mikulinskii 1990; J. McClelland 1983: 180–90).

9. See Greene (1981: esp. chs. 4–6) on the tenuousness of the distinction between Darwinian science and social Darwinism in the late nineteenth century.

3

Reaction and Resistance: The Natural Sciences and the Humanities, 1789–1945

Eric Mielants

We have seen in the previous chapter how secular knowledge slowly broke its ties with, and subordination to, theology in the Western world. We traced especially the ascent of what came to be called "science," a progressively well-defined approach to the study of the world of nature, to the privileged pole in the domain of secular knowledge. However, the structures of knowledge took shape over time as a relationally organized whole. As sets of intellectual disciplines housed in separate departments in institutions of knowledge production, the sciences (unbiased and lawlike) coalesced at one pole of this structure and the humanities (value-laden and chaotic) were consolidated at the opposite pole. This arrangement was not accomplished without resistance. The construction of the dominant position of science was met by fierce opposition from the other camp; furthermore, classification of just what constituted science gave rise to contentious intellectual debate and institutional challenges. From the late eighteenth century this combination of resistance was tagged, at least for a long time, as the "romantic reaction."

The struggle was clearly not unrelated to the upheavals of the French Revolution. At first, both science and philosophy tended to be identified, too readily no doubt, with proponents of the Revolution, but soon new currents of philosophy took hold and seemed to be espoused primarily by its opponents. The terrain was confusing and politicized.

By the end of the eighteenth century, science had come to be defined quite clearly as the search for universal laws in a deterministic and empirically knowable world and thus a world whose past and future could be described exactly, provided one had the knowledge of the appropriate laws and the so-called initial conditions. In the natural sciences this came to be called the "Newtonian worldview." Auguste Comte sought to generalize this worldview beyond the natural sciences, calling it "positivism."[1] Central to the scientific ethos was the disenchantment of nature, a view that justified the conclusion that man could and should dominate nature. In effect, since the seventeenth century, science was doing more than breaking with theology. It was making it plausible for humans to consider themselves godlike in their relation to nature.

Nonetheless, at the end of the eighteenth century, most scientists were not yet ready to state this so baldly. Most still thought (or argued) that religion was compatible with science (von Engelhardt 1988: 118). Nor was the institutionalization of knowledge into distinct disciplines something yet self-evident. Furthermore, the gradual processes of secularization and specialization did not play out uniformly in every country. These struggles over how to conceive of knowledge were taking place, for a long time, primarily in three countries—France, England, and Germany. However, the political situation, as well as the institutional role of the structures of knowledge, was different in each.

It seems sensible to start with France, where the French Revolution had an immediate and major impact on the emerging two-cultures split. In the beginning, conservative thinkers identified philosophy (that is, Enlightenment philosophy) as bearing great responsibility for the excesses of the Revolution, and especially *la Terreur*.[2] Attempts to contain it may have been reinforced by an inherently mystical Christian reaction by a branch of spiritualism within French philosophy (Maine de Biran, Victor Cousin) which sought to create a legitimate, alternative way of thought to that of the Enlightenment (Verdenal 1973: 43–65; Dhombres & Dhombres 1989: 480). In the nineteenth century, statistical laws were being formulated by Pierre-Simon Laplace and Adolphe Quételet (who stressed determinism of the strictest sort while denying the possibility of free will), and the natural sciences cast philosophy completely aside, saying it was speculation and no better than revealed knowledge. Both Laplace's "political mathematics" and Quételet's "social physics" were "designed to remove value considerations from the formulation of policy by transforming political issues into epistemological ones" (Peter Buck 1981: 29).

Literature and the arts also suffered in the intellectual aftermath of the Revolution. As Johan Heilbron says,

> While the natural sciences were perceived as specifically useful disciplines, literature became associated with aristocratic salons and frivolous meetings without practical applicability and significance. The increased status of the natural sciences went hand in hand with a loss of prestige in the humanities, which resulted in a quite sudden turnabout in the existing intellectual hierarchy. (1990: 135–37)

During the Napoleonic period, but also in the ensuing period of the Restoration in France, literature and the arts became dominated by religious, romantic, antirationalistic, and antiscientific figures like François de Chateaubriand, Madame de Staël, Louis de Fontanes, and Louis de Bonald. By their writings, they "wage[d] an open war on the sciences since [in their view] the latter's evolution only brought about moral corruption and atheism" (Heilbron 1990: 164; cf. Dhombres & Dhombres 1989: 278, 326). By 1807, Bonald was writing of the divorce that had taken place between the natural sciences and literature (Heilbron 1990: 166–68; cf. also Dhombres & Dhombres 1989: 430).

As the split between the two cultures widened,[3] the methodological views of the scientific camp spread beyond the natural sciences. Beginning in the second decade of the nineteenth century, the political and "moral" sciences were urged to employ the methodology of the natural sciences (Daston 1988: 287, 371). This pressure, which came from such figures as the early Comte, drew sustenance from the intellectual tradition of the Enlightenment in which eighteenth-century philosophers in general "took the scientific world picture which Newton had given them for granted" (J. Bernal 1971: 516). There was a continuous line from the Enlightenment to Condorcet and Laplace, who both argued that the moral sciences should follow the same method as the physical sciences "in order to attain the precision which distinguished scientific truths from intuitions" (Daston 1981: 291–92). The line continued with Charles Fourier, Saint-Simon, and the early Comte, who wrote in 1824: "The laws governing the development of the human specie are as determinate as those governing a falling stone" (Dhombres & Dhombres 1989: 302, 370, 534).

Views were quite divided. On the one hand, the "social mathematics" of Condorcet and his followers was not readily acceptable in the humanities, where they were accused of

> unjustified or even bizarre assumptions, vastly oversimplified hypotheses, and a pernicious tendency to obscure the characteristic complexity of the moral realm by taking averages. Although there was little consensus as to alternative methods or models, the philosophers of the moral sciences were generally united in rejecting both the problems and the assumptions of the probabilistic model [as formulated by academics like Laplace or Quételet]. (Daston 1988: 377–78)

On the other hand, many saw the humanities as unwilling or unable to provide research with exact and predictable information, whereas the natural sciences claimed they could, and did (as their technological innovations seemed to demonstrate). The so-called humanists complained that they were pushed to the sidelines. Bonald, expressing his radical opposition to the introduction of the method of natural sciences into the moral sciences, lamented the "forthcoming fall of 'the republic of letters' and the universal dominance of the exact natural sciences" (Heilbron 1990: 168–71; cf. Dhombres & Dhombres 1989: 613–14). For Bonald, the "superiority of the moral sciences to the natural sciences was manifest," but in the larger French society a shift in prestige ranking was taking place as the "exact sciences were now accounted the superior (hautes) sciences" (Lepenies 1988: 10, 9).

The revolutionary period was one of structural transformation, organizational centralization, and overall institutional growth for the educational complex. Universities and royal academies had been closed and new institutions such as the École Normale Supérieure were founded, as well as institutions specifically geared to the natural sciences and engineering such as the École Polytechnique (Lutun 1993; Markov 1989: 131), the Bureau des Longitudes, and the École de Pharmacie (Verger 1986: 262). By 1808, faculties of letters, science, medicine, technology, and law had been established in Paris and several provincial cities, with no links among themselves, having relations only with the Ministry of Education (R. M. Gascoigne 1987: 322–30). In the Institut de France, which replaced the academies in 1795, the natural sciences, literature and arts, and the social sciences were officially recognized as belonging to different intellectual spheres (Ponteil 1966: 138–40; Heilbron 1990: 121). The "parcellization of scientific knowledge by specialization" is also indicated by the growing number of specialized journals between 1789 and 1825 such as the *Journal de galvanisme, de vaccine, etc.* (1803) or the *Annales de mathématiques pures et appliquées* (1810) (Dhombres & Dhombres 1989: 480–82). The split between the two cultures was thus beginning to be institutionalized during this period in France (Crosland 1975: 40).

The conservative romantic writers, lamenting the split, found themselves unable to turn the intellectual clock back during the Restoration, nor could they undo the educational reforms implemented in 1789–1815. While the humanities in France were now dominated by the (aristocratic) romantic movement, the (natural) sciences slowly became an intrinsic part of bourgeois culture, which in turn was constructed as an alternative to the aesthetic classical culture of the aristocracy (Harrison 1999: 64–66).

The fact that within France the specific form that the Enlightenment took was so "unquestionably central" and that the Newtonian tradition "reigned" (Knight 1986: 26; Halls 1965: 6) meant that no romantic reaction could occur within the natural sciences themselves (Dhombres & Dhombres 1989: 452; Knight 1970: 57). However, if one turns to England or the Germanies, the picture is quite different.

In England the romantic reaction against the specific form of rationalism that excluded values from analytic thought is intimately linked with the "national" war against France. The romantic-conservative-traditional stand (associated with the "loyalism" that originated in a period of determined prosecution of Jacobinism) was part of the historic reaction to an ideological opponent (France), which had become associated with rationalism, revolution, equality, and mass democracy (Klancher 1989: 464). English Baconian induction/experimentalism had always been up to a point hostile to hypotheses, schemes, and prescriptive constitutions. However, to some extent this tonality was modified by the rise of utilitarianism—Jeremy Bentham, David Ricardo, and John Stuart Mill (J. Bernal 1971: 545). The Reform Act of 1832, conceived in the spirit of utilitarianism, opened up a period of reformism, after which utilitarianism itself became "penetrated by positivism" (Duchesneau 1973: 123–50; Kremer-Marietti 1982).

Yet, throughout the nineteenth century, science remained affected by "a powerful culture of hard-headed, no-nonsense Anglo-Saxon empiricism" (Harris 1994: 224). Although Victorian scientists "appreciated the French confidence in mathematics and quantification, they rejected the linear, quasi-mechanical French view of progress," preferring a biological metaphor implying that

> change is imperceptible and slow, and [that,] in contrast to the rectilinear Turgot-Condorcet model, the tempo cannot be altered. Revolutions alter tempo. Evolutionary rather than revolutionary views resonated with the dominant British philosophical and political position regarding continuity. (Schweber 1985: 21–22)

In nineteenth-century England, the Royal Institution (founded in 1799) would become a "successful venture in balancing what came to be called pure and applied science for the benefit of the larger public" (Nye 1996: 25), while the British Association for the Advancement of Science (BAAS, founded in 1831) was more important than the Royal Society, which was dominated up to 1860 by nonscientists who were more interested in social eminence than in scientific research (R. M. Gascoigne 1987: 387; Lyons 1968: 228–71). Shortly after the founding of the BAAS, William Whewell, a quintessential philosopher and scientist-mathematician, thought it was time to invent a neologism, "scientist," in order to make the distinction between a philosopher and a scientist—an indication of the strength of the imperative to specialize in England.[4]

As nonscientists gradually became a minority in the Royal Society, "Victorian science saw the development of specialized disciplines, pursued by paid professionals who were of necessity minutely educated in their fields" (Merrill 1989: 75). For instance, both Alfred Wallace (b. 1823) and Thomas Huxley (b. 1825), who received only minimal formal schooling and were largely "self-educated and self-taught in science," were nevertheless capable of acquiring broad scientific recognition in the 1850s (R. M. Gascoigne 1987: 395). While this was still possible (but more difficult) a generation later (e.g., Oliver Heaviside [1850–1925]), it became rare that one would "think oneself" into a science (such as mathematics or chemistry), let alone end up as a professor in a leading institution. The number of "pragmatic men" who were "craftsmen transmuted into professional experimental investigators [and who] brought the attitude of the craftsmen in the research laboratory" (Crowther 1966: 77–78) decreased after 1850, and they had disappeared almost completely by the end of the century.

The evolution of the *Philosophical Magazine* (!) illustrates the increasing specialization that was a *conditio sine qua non* for the two-cultures split occurring between 1830 and 1860:

> Though by the end of the nineteenth century it was one of the best known and most highly respected journals in the scientific world, it came originally from humble origins . . . originally cover[ing] the whole of science it gradually became restricted to the physical sciences and by the end of the century very largely to physics. Such specialization was essential for the survival of a non-institutional (or proprietary) journal in the late nineteenth century. (R. M. Gascoigne 1987: 387)

The same trend of specialization was also reflected in other journals and reviews such as *The Zoologist* and *Nature*. In the period between 1850 and 1880,

> the gradual elevation of the level of discourse up to "professional" standards made it more and more difficult for the interested "amateur" or the general public to share the intellectual context of such discussions, [making scientists] less and less used to understanding and discussing science within a broad, accessible, public context. (Roos 1985: 159–80)

The gradual removal of both the pragmatic men and the noblemen from the scientific community after 1850, what has been called the "exclusion of the amateur" (Merrill 1989: 79), had of course much to do with the reallocation of resources toward a new locus of knowledge, the new university, which formalized education through certification and acquired an almost complete monopoly over the production and dissemination of specialized knowledge by the end of the nineteenth century.

The reforms initiated by Wilhelm von Humboldt in Germany (Rüger & Klein 1985; C. McClelland 1980; Van de Graaff 1973), and the *grandes écoles* founded by Napoleon were both examples of the efforts of new state-sponsored institutions in the early nineteenth century publicly to promote—and guarantee the possibility of—the specialization of academic

scholarship. Whereas up to the late eighteenth century intellectual specialization had predominantly been dependent on private patronage, the
nineteenth-century reforms highlighted the responsibility of the state to
provide for and guarantee the required production of (applied) knowledge. The university structures in which the unity of research, teaching,
and learning was established were increasingly specialized (Nye 1996: 8;
Jungnickel 1979: 41). While the function of eighteenth-century universities had been the dissemination of knowledge, rather than its production, and private or official societies were at the forefront of new
discoveries (Roberts 1991: 244–45, 371), the early nineteenth-century reforms enabled the universities gradually to take center stage again. This
was especially the case in Germany, where universities replaced the academies as the home of scientific research by 1830 (Steven Turner 1971:
137–38). By the mid-nineteenth century, the creation of seminars desig-
nated by "disciplines" led to the emergence of a "professorate with primary loyalties to a given discipline rather than a particular institution"
(Dauben 1981: 381; Steven Turner 1971: 145). Consequently, the require-
ments on academics to do research and publish became very apparent
by the 1840s. A distinction between "contributions to science" and "contributions to practice" was generated, and this was a factor in the forging of academic communities in several of the disciplines.

   This restructuring of national higher education in Europe caused a
major debate as to how this should be implemented, that is, what should
be the division of labor and consequently to whom funds should be
allocated. This cannot be analyzed separately from the two-cultures split.
Natural science argued that it could deliver major technical innovations
and applied scientific contributions. The arts and humanities, also located within the new university structure, found themselves in difficulty in
attempting to respond to the claims of the new, successful methodology
of the natural sciences. Scientists sought to differentiate themselves from
philosophy, using the fact that "the successes of experimental science
seemed to be proving that it at least could do quite well without traditional philosophy. . . . Scientists simply began to shy away from philosophy and retreat into experimentation" (Gregory 1977: 7). Some were
more aggressive. Thomas Peacock "attacked poetry as an infantile and
archaic holdover from primitive times that interfered with serious thinking and that was bound to degenerate because it did not address itself to
the scientific and philosophical part of the community" (Korg 1985: 141).

   These attacks, of course, had a direct impact on the status of philosophy in the hierarchy of scientific institutions. Yet, presenting the historical evolution of new institutional settings as a unilinear process that
resulted in continuous scientific specializations in all countries would be
erroneous. In France, the Revolution led the state to invest massively in
science (Fayet 1960: 474; Rudy 1984: 127), which meant that science became a profession, whereas "under the *Ancien Régime* the man of science

had hardly been distinguished from the man of letters [and] the terms savant or philosophe could be applied to both" (Crosland 1976: 141; Dhombres & Dhombres 1989: 611). However, by the 1820s the German university system was quickly catching up, thanks to better staffing and financing (Shinn 1979: 291); and in contrast to France, where the Revolution opened up an intellectual space for militant atheist scientists and where "religious explanations were separated from scientific explanations" (Dhombres & Dhombres 1989: 256, 432), the link between science and religion in nineteenth-century England indicates the lateness of the professionalization of science there (Knight 1986: 17).

The French situation thus differed sharply from that of Great Britain,

> a society where science was still a part of the culture of an educated middle class rather than a specialized activity offering full-time employment and remuneration [and where science] was done in the setting of literary and philosophical societies, where science was still formally associated with literature as polite culture rather than a specialized intellectual activity. (Crosland 1976: 154)

The British idea of self-help for science (Knight 1986: 132, 137) paralleled its relatively late legislation regarding universal elementary education (den Boer 1998: 177).

Indeed, in Prussia education was compulsory for all children as early as 1717, which is why by 1830 it was called "the land of schools and barracks" (Farrar 1976: 180). Thus, in Germany, the state was from early on engaged in supporting both education and science (Steven Turner 1981: 110; Diebolt 1995: 100), which it institutionalized from the top. Nevertheless, the link between philosophy and science made the concept of two distinct cultures dubious in early nineteenth-century Germany (Knight 1976: 166). Mathematics and the natural sciences only "embarked upon the preemption phase between 1825 and 1845" (Steven Turner 1981: 112), and within the faculty of philosophy, the natural sciences witnessed a steep rise in disciplinary research units only after 1860 (Lundgreen 1983: 171; C. McClelland 1983: 307). Specialization increased tremendously, however, in the second half of the nineteenth century. By 1850, the career scientist was practically always a professor, while the laboratory in which he worked was a part of the university structure (Steven Turner 1971: 137). The natural sciences at the universities were elevated "from the position of mere preparatory subjects auxiliary to practical professions to that of autonomous academic disciplines that trained their specialists in their own fields" (Jungnickel 1979: 15).

In England, despite its Newtonian tradition, the two cultures received an institutional base only in the second half of the nineteenth century (Knight 1986: 137) when the state was finally willing, as a result not only of longtime complaints about "the decline of science" (for example, by Charles Babbage in 1830 and by Huxley in the 1860s) and laments about "traditional modes of patronage" (MacLeod 1981: 26), but also because of the international situation, to make up for the "long time lacking of organized social and state support" for science (Manten 1980: 14).[5] Great Britain was characterized by individual scientists who were among the well-to-do or patronized by the well-to-do. In the absence of an elaborated network of state-sponsored institutionalized settings for the production and dissemination of knowledge, "the social structures of shared education, training, and research" (Morus 1992: 23) that enabled qualified experts to specialize in a milieu wherein professional (academic) careers were systematically nurtured and guaranteed by the state remained virtually nonexistent in the first half of the nineteenth century.[6] And when "belatedly, the [British] Government began to give a measure of support, the sums contributed could not compare with the lavish state expenditure in Germany" (Roderick & Stephens 1972: 12).

Yet, depicting intellectual history as the inevitable unfolding of an unavoidable two-cultures split as if it were a natural evolution due to increased state competition would be incorrect. The romantic movement between the end of the eighteenth century and the mid-nineteenth century generated an alternative method of inquiry within the natural sciences based on imagination and personal experience.

Indeed, constructing a rigid dichotomy between romantic literature/ arts and the scientific method of inquiry of the natural sciences is somewhat misleading (J. Smith 1994: 90–91). Why does art have to be labeled personal, subjective, spiritual, emotional, and organic, while science is labeled impersonal, objective, physical, logical, and mechanistic (J. Smith 1994: 45)? It is precisely the "rationalist discourse of science [that] required and created the cultural fiction of just such an antithesis between artistic subjectivity and scientific objectivity in order to validate its own absolute claim to the truth" (Black 1990: 133). Throughout the early nineteenth century, romantics were very much occupied with theory and science (e.g., the Frankenstein novel), but they mainly turned against "Newtonian" science. Keats called it "cold philosophy" (J. Smith 1994: 46). They also generally "disliked seventeenth- and eighteenth-century natural history for the passivity of mind that it implied: classification was a sterile, unimaginative exercise in pigeonholing that took no interest in the relationship between observer and observed or between the object and its environment" (J. Smith 1994: 60).

Their main charge, however, was against materialism and crude Newtonianism rather than against scientific inquiry per se. For example, William Blake in 1820 praised the "wedding of art and science" (Burwick 1986: 8), and, conversely, some nineteenth-century scientists were also much concerned with art and the "position of the aesthetic." Needless to say, not everyone in the arts and science was in favor of a rapprochement. The reaction of some theologians and philosophers alike toward science was sometimes scornful as if "the results of natural science were trivial and relatively unimportant" (Preyer 1985: 51), which may have reinforced and perhaps even quickened the pace of the emerging split between the two cultures in the late Victorian era. In any case, the romantic movement ultimately was incapable of providing a "scientific" alternative and could not graft itself onto the scientific culture.

The final outcome aside, it was in the Germanies that the romantic reaction "against the reduction of the natural world to mechanisms and mathematics" was strongest, and it lived longer there than elsewhere (Nye 1996: 25).[7] Here the romantic movement, embracing a "philosophy of nature, which favored unification of the various branches of physics over a utilitarian approach to research," became particularly important in science (Bevilacqua 1994: 21). For instance, Johann Wolfgang von Goethe's attack on Isaac Newton's color theory was launched at "a time when the possibility of universal knowledge, mastery of the arts and sciences, still seemed open to the ambitious mind" (Burwick 1986: 3). When his *Zür Farbenlehre*, methodologically "grounded on empirical psychology and, ultimately, on a physiological anthropology" (Bell 1994: 289), was published in 1810, physicists accused Goethe of "dilettante speculation," while literary critics and art historians chose mostly to ignore this study (Burwick 1986: 9).[8] "[N]o physicist or chemist of note [was] able to accept his views. On the other hand, he was always supported by men like Hegel, Schelling, and Schopenhauer, who were not primarily scientists" (R. Gray 1952: 129); he was also supported by the "self-made chemist" Johann Wilhelm Ritter and the poet Friedrich Schiller (Snelders 1994: 37). For many nineteenth-century romantics, and notably for Goethe, Newton was "the ideal target, because he represented a hated ensemble of ideas: the reduction of phenomena to mathematics and (according to Goethe) the use of contrived experimental environments" (Bell 1994: 3).[9] Newtonian science made "the motions of the predictable with a precision limited only by the inaccuracies of our measuring instruments; there seemed every reason to believe that this held true for all forms of matter. In this fully deterministic model, all effects had definite causes, ascribable to known laws" (Tolstoy 1990: 20).

In *Zür Farbenlehre*, Goethe "insisted on involving the observer in the phenomena observed." Just as in his botanic study *Metamorphosis on Plants*, his approach "involved the observer in a very subjective, intuitive way, allowing him the full range of sense impressions in the study of a plant" (Williams & Steffens 1978: 18–19). Not making a distinction between "subjective" and "objective" nature (Snelders 1994: 39) was obviously a sin against one of the Newtonian scientific premises, which "excluded the observer by relying on exacting observational techniques and mathematics. This exclusion was unacceptable to Goethe since it was unnatural; man was part of nature and should not be removed from it" (Williams & Steffens 1978: 20).

Romantics, who were preoccupied with theory and science alike, balked at a specific (Newtonian) form of science that, through its deterministic and (they believed) sterile outlook on nature and the world, "made inroads upon their imagery" (Merrill 1989: 7), much as they were discontented with the rationality of an Enlightenment of which "science was to be its tool and Newton its prophet" (Tolstoy 1990: 173). The opposition of the romantic reaction to the Aufklärung in Germany made for a major collision. While the latter advocated the explanation of natural phenomena in a Newtonian mechanistic and atomistic way, the former protested against the reduction of all natural science to mechanics (Knight 1976: 169), proposing instead dynamic and organic formulations in which the unity of all natural forces was central (Snelders 1994: 46–47). They thereby rendered the examination of material particles unimportant or even imaginary (Knight 1976: 164–65).

Romantics such as Friedrich Hölderlin and Novalis faced a common predicament: "the tyranny of philosophical and scientific systems based on the opposition between world and mind, between outward and inward reality" (Hamburger 1957: 72).[10] The investigation of nature through original intuition such as in Schelling's *Naturphilosophie* and mysticism (Gower 1973: 301–56) is indicative of how a romantic alternative came into existence, while professional natural scientists were creating a culture of their own. The latter later on depicted Goethe as a "tragic" example of how amateurs should stay away from "hard," exact, and predictable science.[11]

From the 1790s to the 1830s, the romantic notions of a "dynamic" nature that was unified and "hylozoistic" dominated both the German philosophical realm and the natural sciences (Snelders 1994: 16). Humanity and nature were believed to "share a common destiny" (von Engelhardt 1988: 115), and the unity of knowledge was still being emphasized. Many early nineteenth-century German scientists rejected "the godless, mechanistic science of the late eighteenth century in favor of a [worldview that] included human freedom and moral choice" (Culotta 1974: 5–6). German natural scientists were open to speculation, deductive methods, intuition, and the interpretation of natural phenomena through an *innere Gefühl* (Snelders 1994: 21). In the second decade of the nineteenth century, debates still raged about the *vis vitalis* in organic chemistry (Markov 1989: 135).

To be sure, Naturphilosophie had its proponents in Great Britain (such as Richard Owen and Michael Faraday); nevertheless, most English scientists followed the Newtonian tradition and attempted to explain natural phenomena using mechanistic models (Snelders 1994: 153). All the same, to call the romantic alternative an "anti-scientific movement" (Bevilacqua 1994: 21) is an exaggeration. Before 1850 the cleavage between the pragmatic men and pure academic professionalism was still limited, and here the proffered alternative to science was not antiscientific but rather a "moral revolt." It attempted to "defend qualitative, non-mathematical science: science which draws the scientist into it as an involved observer. For the Romantic, science should involve subjectivity and an individual, intuitive response to nature observed [in] the study of natural history" (Williams & Steffens 1978: 15–16). Herder also formulated an alternative perspective, which came out of his romantic preoccupation with the *Volksgeist* and the uniqueness of nations/peoples (Rupke 1983: 390, 407), by stating that if "natural language is taken as the paradigm of reason, instead of some explicitly universal structure of thought such as mathematics, this entails that the scope of reason is now culturally bound and hence incapable of delivering universal truth, but only truth relative to culture" (Grier 1990: 234–35). As Olson sums up, in disagreeing with a particular form of scientific inquiry (the Newtonian), most nineteenth-century romantics criticized both the French and the Industrial Revolutions in various ways.

> [Since they were] so closely linked with scientific thought, it was natural that their negative features were blamed by some on the extension of scientific thinking into inappropriate domains. . . . Against the authority of reason, the obsession with quantification, the worship of universal truth, the assumption of analyticity, and the optimism with respect to individual's ability to formulate intelligent rational courses of action, the new intellectual leaders of reaction offered the authority either of tradition or of the poetic imagination. They almost uniformly denied that only what can be counted counts. They insisted that wholes are more than the aggregates of their parts. They valued the unique over the universal, the distinctive over the regular, the divine over the natural. . . . Blake and Wordsworth had remarkably similar responses to both the French and Industrial Revolutions and to scientific liberalism [as they] were appalled by the coldly rational, quantitative, and managerial mentality. (Olson 1990: 368–69)

As a result, from the time of Matthew Arnold to that of C. P. Snow, the critics of the revolutions (which they linked with the Newtonian scientific method) were classified as "the right wing, and the scientists, as befits a radical movement, the left" (Connell 1971: 187). Adherents of the right wing were identified as the traditionalists, the romantics, the classicists, the "non-" or even antiscientists.

In the sciences, it took time before the bridge between the realm of practice (rule of thumb) and the realm of pure science was no longer solid enough for anyone to venture across. The pragmatic men such as Sir Humphry Davy and Faraday (M. Davies 1947: 172) remained more open to the more experimental and speculative elements within Naturphilosophie (as embraced by the romantics), less hostile than the Newtonian scientists-professionals who were trained within the academic world, which after the 1850s was increasingly dominated by them. By

1850 the enhanced social and cultural status of the natural sciences enabled them to "liberate themselves from their previous subordinate existence to the arts and medicine." The founding of independent science faculties "was reflected in the adoption and adaptation of specialist terms within the common core of discourse and in the acceptance of scientific terminology in other disciplines" (Townson 1992: 61).

By the end of the nineteenth century little intellectual leverage was left for a romantic alternative in the academic world. At a time when more walls were being erected with the increasing compartmentalization of knowledge, the romantics were unable to obtain a lasting niche for themselves in the newly emerging science faculties. Although the romantic alternative did not disappear completely after 1850, in the heyday of positivism it was reduced to an irrelevant, insignificant undercurrent. In France and Britain the "empirical" natural sciences gradually came to preempt the field. In Germany a real "science war" did take place between those who advocated a romantic alternative and those who propagated an empirical-positivistic approach, but the ultimate triumph of the latter resulted in a general attempt to "purify" the natural sciences of all metaphysical undertones by men such as Jacob Moleschott, Ludwig Büchner, and Ernst Haeckel. Only at the very end of the nineteenth century did some scholars, such as the physician Ernst Mach and the chemist Wilhelm Ostwald, attempt to reintroduce a connection between natural science and philosophy in Germany (Snelders 1994: 184), while in late nineteenth-century France Henri Bergson broke with Cartesian and Kantian traditions to attack determinism and materialist mechanism.[12] In the England of 1914, Bertrand Russell advocated that philosophy should draw inspiration from "science" rather than from "ethics" and "religion," thus casting aside concepts such as good and evil.[13] It was only in the interwar years that such scholars as Alfred N. Whitehead and L. Susan Stebbing prepared the ground for later criticism.[14]

Humanists only grudgingly accepted the separation from the natural sciences within the structures of knowledge, and the implicit hierarchy was increasingly felt in departmental budgets and social prestige. While some humanists attempted to bring positivist methods into the humanities in order to save them from the accusation of irrelevancy in a society more and more obsessed with the functionalist efficiency advocated by John Stuart Mill (Dale 1989: 14), those who continued to fulminate against scientific management and technocracy maintained a firm belief in the superiority of the humanities.

It must be noted that the humanities were not immune to the influences of the "cultural revolution" that had taken place by 1900 (N. Stone 1999: 295–304). The mathematization of the humanities took off in the latter half of the nineteenth century: mathematical patterns in painting, architecture, music (J. Davis 1995), as well as mathematical approaches to philosophy and structuralism in literary criticism, are all examples. Positivistic ideology had by then "infiltrated all aspects of thought, from art to science to philosophy" (Rollin 1998: 66). While the tertiary sector expanded in every state, the more highly educated middle class increasingly snubbed nonfunctional and nonutilitarian disciplines (e.g., classical languages) in the educational system (Gerbod 1992: 299–300). Although these subjects were mainly defended by an elite, they performed a most important function of propagating high culture as the salvation of the nation.

From the mid-nineteenth century on, the teaching of a "national" literature became central in forging a national identity. English literature was part of the examination of civil service candidates (Baldick 1983: 61). But what appears crucial in the creation of English studies, scorned by the more traditional humanists as the "poor man's Classics" (Eagleton 1983: 27), was the cherished idea in the higher educational system that a unifying national identity could bring about the reconciliation of all classes (Court 1992: 41). The tumultuous Luddite uprisings and the Chartist movements, which painfully exposed socioeconomic strife between the "haves" and the "have-nots" in British society, encouraged the middle class (especially the utilitarians) to espouse the idea that literature and the study of English should be used to promote cultural unity and political stability (Court 1992: 94; Palmer 1965: 42). Thus, from the 1830s onward English studies and English literature were enlisted as essential tools to "unite the nation and prevent class conflict" (A. Bacon 1986: 610; Eagleton 1983: 27).[15] By the mid-nineteenth century the institutionalization of English literature was being supported both by the classicists who "saw it as an additional weapon in the armory of culture to use in the battle to halt the advancing forces of science in education" (A. Bacon 1998: 12) and by the scientists themselves, who were convinced that English should replace the obsolete classics as an educational instrument. Eventually, the study of English was institutionalized in even the most elitist universities and colleges. This role of English literature received official sanction in the "Newbolt report" of 1921 (Samson 1992: 13; A. Bacon 1986: 611). Ironically, the particular political use of English in a nationalist and racist framework (Parrinder 1991: 170) was concealed in the twentieth century by a methodology applied to the study of literature that isolated texts from history and context and claimed to pursue "pure, disinterested knowledge" (Barry 1995: 15, 26). It allowed academics to maintain, well into the 1950s, that their discipline was essentially an "apolitical" one (Court 1992: 116).

In France and Germany, courses in their national languages and literatures had been institutionalized earlier than in England (Court 1992: 116). In France in particular, the government of the French Revolution attempted to implement linguistic policies that would eradicate local languages and impose a unified language in the entire country (Bodé

1996: 779). A series of more tentative reforms in the period 1815–1870 attempted to diminish the importance of classical languages in secondary education in favor of instruction in French (Moody 1978: 32, 39).

The petite bourgeoisie in particular, in embracing the importance of scientific activities (Harrison 1999: 49–86), provided the backing for educational reforms, which resulted in the creation of the *écoles primaires supérieures* in the 1840s and "special education" in the 1860s to include the teaching of French and commercial courses at the expense of the classics (increasingly reserved for a tiny elite, especially those pursuing higher education in the university system) (Gildea 1980). Although in the period 1815–1870 the state gradually increased its responsibility in the (elementary) education of French citizens (Furet & Ozouf 1982: 131), it was only in the wake of a growing Prussian menace after 1866 that the government "gave new impetus to efforts to eliminate linguistic diversity" (Moody 1978: 77; Furet & Ozouf 1982: 282). Ultimately the defeat of 1870 was the major turning point that lent legitimacy to the implementation of a series of systematic policies to turn "peasants into Frenchmen" (E. Weber 1976). After 1870 the teaching of French and the use of music became quite central to the unification of the nation (Alten 1996).[16] Shaken by the trauma of 1870, the country embarked upon an almost unanimously approved policy of *francisation* (Ozouf 1985: 65), which gained speed in the 1880s when a free and compulsory elementary school system and a more accessible high school system were provided by the state for its citizens (Halls 1965: 22).

As in England, the formation of a "general culture" remained the cultivation of "general moral and aesthetic principles" (Talbott 1969: 13), and as in England, classicists were divided as to what extent they should welcome the study of modern languages as an ally against science or oppose it on the grounds that modern languages were not part of a cultural education at all (Prost 1977: 255; Talbott 1969: 14). Yet as the sciences became after 1850 a larger part of a higher education at the expense of the humanities (Moody 1978: 62–63), the classics very gradually lost ground to French and to the modern languages in general (Delesalle & Chevalier 1986: 229; Prost 1977: 78, 247). Although the classics remained important as marks of social status for the upper class (Gildea 1980: 293), the bourgeoisie both in France and in Germany saw them more and more as an aristocratic snobbism, even a danger to the national wellbeing (Talbott 1969: 15–16; Townson 1992: 57). The Commune as well as the social strife around the secularization of the public school system in the 1880s also exemplified the urgent need for national unity, class cohesion, civic responsibility, and "devotion to the nation" (Moody 1978: 89), which of course had to be achieved in French, not in a classical language. The "modern" humanities were essentially assigned the task to civilize and to construct a national identity,[17] while specialization of the sciences encapsulated all the hopes of revenge on Germany (Fox 1984: 108–9).

In the Germanies, the reaction to the Napoleonic occupation led to an emphasis on the concept of a *Kulturnation,* a nation "based on a common cultural heritage," in which language became an "indispensable tool" (Coulmas 1995: 57). Beginning with Johann Gottlieb Fichte's influential "Addresses to the German Nation" in the early nineteenth century, language and nationhood became ever more intertwined. It was also in reaction to Napoleon's initial success that states such as Prussia took steps to reform their educational system. By founding *Volksschulen* that would "cultivate German nationality," state officials sought, rather successfully, to "create a nation from above" (LaVopa 1979: 434, 447). Throughout most of the nineteenth century, the German higher educational system obtained much more state support than that of any other country. The bureaucracies of the states (and after 1871 of the empire) supervised the universities firmly (Ringer 1967: 124), in stark contrast to France, where reform in the university system was brought about primarily because of increased competition from the *grandes écoles*. If the German romantics had failed in their quest to influence the natural sciences after the 1840s, they were very successful in channeling their energies to serve the political movement of national unification. Indeed, the construction of a German "national science,"—*Germanistik*—in which the romantics Jacob and Wilhelm Grimm played a central role, was a fusion of the humanistic disciplines (the study of German history, philology, law, literature, and language) to "strengthen national consciousness" (Sauer & Glück 1995: 76). Thus, the desire to study and teach German literature became in reality inseparable from a political and social involvement in the unification of the Germanies (Hohendahl 1989: 107, 142). After the failed revolution of 1848, German studies gradually turned into a "pillar of Prussian conservatism in the service of the state establishment" (Townson 1992: 95). By the 1850s literary history had become "an academic discipline with its own academic chairs." But it was not until the 1870s, when political unification had finally been accomplished, that it moved from addressing the general public sphere to becoming a genuine academic discipline of its own (with an often heavy emphasis on linguistic purism) and attempted to separate itself from "mere" journalistic criticism (Hohendahl 1989: 201, 242). This was done by emulating the methods of both classical philology and the "exact sciences" (Townson 1992: 95). As in France and England, the study of the nation's language in post-1870 Germany was deemed essential to the education of the youth "towards a spiritual, determined and joyful German-ness" (Hohendahl 1989: 97) and producing civil servants for the expansion of the colonial empire.

In sum, despite the humanists' protests against the presumptuous rise of the natural sciences, the "aesthetic background to European education was crumbling" (N. Stone 1999: 296). Most propagators of high culture put themselves at the service of warfare and nationalism in 1914–

1918 (Roshwald & Stites 1999; Townson 1992: 104), and by the 1920s at the latest, the "hard sciences were seen as the dominant field of inquiry due mostly to their rigid methodology and the conclusiveness of their findings" (J. Davis 1995: 506).

In the context of this process, the mutually exclusive value systems that were associated with humanist social criticism, and their manifestation in the specific political projects of radicals and conservatives during the nineteenth century, determined the intellectual definition and institutional emergence of the social sciences. The triumph of the empirical-positivist methodology in the natural sciences throughout the core of the world-system informed the constitution of the social sciences as disciplines producing value-free knowledge of human reality and thereby paving the way for a "new liberal" consensus (incremental, "scientifically" directed "progress"), a consensus that papered over the ideological split. By definition, the social sciences appeared to be more open than the humanities in the debates over the extent to which the methodology of the sciences should be copied.

Nonetheless, the "historical economic school" (*Staatswissenschaften*) in the Germanies resisted the takeover by deductive methods as well as the mathematization of economics much longer than was the case elsewhere (Barkai 1996). Furthermore, in the 1880s, Wilhelm Dilthey emphasized the importance of empathy and *Verstehen* of historical empirical reality and condemned "the abstract school" of positivism and the infiltration of methods of the natural sciences into history. Although Dilthey wanted history to be labeled "scientific," he proposed to incorporate "the good and the beautiful" in a historical "search for the truth" in his *Introduction to the Human Sciences* (Makkreel & Rodi 1989). This was in sharp contrast to more "positivistic" historians such as Karl Lamprecht and Friedrich Meinecke in Germany or Thomas Buckle, John Bury, and Frederic Harrison in England (MacLean 1988; N. Wilson 1999). Yet, by the turn of the twentieth century, the "standard view" in the philosophy of the social sciences stressed

> the unity of natural and social science in opposition to more speculative forms of social theory, the importance of empirical testability and the value-freedom of social science. . . . The mission of social science was to imitate the natural sciences in the accumulation and integration of factual knowledge and law-like generalizations. Work which did not do this, by implication, was idiosyncratic, impressionistic, novelistic and so forth. (Outhwaite 1996: 84, 93)

For many, the nineteenth-century "sociology of change" was embedded in a belief that science would be able to discover societal laws, thus enabling sociologists—conservative and progressive alike—to turn their profession into a form of "social engineering" (Bramson 1974: 23).

As the nineteenth century drew to a close, the "ivory tower of art furnished the only solace and refuge" (Randall 1976: 592) for those of nonscientific temperament. It was argued that the theories of "Romantic" philosophers like Schopenhauer were a "viciously malevolent, existence-hating attempt to uphold volition in the name of whim worship, while the esthetic Romanticists were groping blindly to uphold volition in the name of man's life and values on earth" (Rand 1969: 90). The humanities became the last locus in the university where "moral values" could be upheld and aestheticism considered legitimate in its own right. When the First World War "marked the end of the great era of Romanticism" (Rand 1969: 107), the gap between the two cultures was complete. Values had been banished from the natural sciences, whereas they had become the defining core of the humanities and the arts, constituting "the conscience of mankind." As Philo Buck Jr. put it clearly in 1930:

> There is no intercourse between the world of art and poetry and the world of reason and demonstration: an armed truce that more than once gives place to active hostilities. The scientist charging that poetry and art are essentially unreal, and the pastime of utterly irresponsible persons with no conscience for fact and law; the poets responding, many of them, by going over in panic into the camp of art for art's sake and subjectivity, and proclaiming the utter independence of art of any practical considerations. It is not quite an edifying spectacle. . . . Science and poetry are inveterate enemies. (Buck 1930: 221–22)

Proponents of the humanities did continue to assert their importance by stressing their functional tasks in society. The construction of nationhood and the pursuit of a civilizing process in which the "cultivation of intellect and sensibility" (Woodhouse [1959] 1968: 55), as well as the mission "to heal a decadent human condition" (Morgan 1996: 317), were central themes. Despite the fact that many people still had much respect for the civilizing project of the humanities in which the *hommes de belles lettres* were the core, "the conscience of a generation" (Rietbergen 1998: 415), "literature ceased to have the central role it had enjoyed before 1914" in the educational system (N. Stone 1999: 306). When in the late 1950s C. P. Snow initiated yet another debate about the two cultures, reminiscent of that between Matthew Arnold and Thomas Huxley (Lepenies 1988: 155–81), Snow, a nuclear physicist and top bureaucrat, could already be considered "the embodiment of the fusion between the powerful exact sciences and the state" (Rietbergen 1998: 426). From Goethe and Schiller, to expressionism and the art for art's sake movement, to Arnold and F. R. Leavis, a direct line can be traced (Egan 1921; Bell-Villada 1996: 290; Morgan 1996: 326–39), a line that stood for the defense of values and beauty in the "long nineteenth century . . . the age of triumphant positivism and materialism [as] science, rather than religion, was on its way to becoming the new faith" (Rietbergen 1998: 402–3). The bitterness of humanists, who from the mid-nineteenth century onward fruitlessly attempted to ridicule scientific achievements (Lalouette 1998: 826) and science's utilitarian claims (Hacking 1983: 472–73), can well be illustrated by the fact that Arnold "wrote to a friend he would rather have his son believe the sun went around the earth than that his mind should be overwhelmed by the natural sciences" (Lepenies 1988: 158). By the late nineteenth century, Arnold had to admit that poetry had been turned into an appendage of science (Benson 1985: 316). By 1945, for humanists such as Leavis the "technological-Benthamite world outside [the university] and the 'life' preserved within the beleaguered, chalk-dusted walls of the university English departments" (Young 1996: 13) had become absolute dichotomies. As the natural sciences and the social sciences of a positivistic stamp increased in prestige, the residual authority of the moral sciences hung on to their mission of "providing moral orientation for mankind" (Lepenies 1988: 11) and preserving the nation's artistic sense of beauty. Literature thus was espoused as the cure to "the soul-destroying evils of a rapidly changing society" (Lepenies 1988: 175), which was dominated by the hard sciences, whose purpose was, as Samuel Taylor Coleridge had put it already in the early nineteenth century, "the communication of truths—truth absolute and demonstrable" (Buck 1930: 216). As both values and aesthetics were banished from the realm of the natural sciences, the former became located exclusively in the humanities, where a profoundly defensive position was maintained up until the rebellion of the 1960s, which, as Arthur Mitzman points out, cannot be separated from "the social romantic critique of nineteenth-century liberalism and [its] anti-Cartesian philosophical sources" (1996: 682).

Notes

1. Comte's positivism "disassociated itself from everything which is non-verifiable, i.e. everything that is not derivable from sense impressions" (Loen 1967: 120–21). Positivism attempted to "clearly demarcate science from non-science, to eliminate metaphysics from science and lay it to rest as 'nonsense'" (Rollin 1998: 68).

2. Condorcet, for example, was an active member of the National Assembly. Conservatives such as de Maistre and Barruel "assumed that the Revolution was the natural product of the Enlightenment and totally repudiated both" (Hampson 1990: 266).

3. Pointedly illustrated by two competing maxims: Benjamin Constant, author and friend of Madame de Staël, was probably the first to use the expression "art for art's sake" in 1804, while the physicist Jean-Baptiste Biot, a pupil of Laplace, introduced the statement "nothing is more beautiful than truth."

4. Nonetheless, although Whewell himself regretted the passing of an "age in which men could move freely between the disciplines of science, philosophy and theology" (Yeo 1981: 69), the BAAS chose to exclude philology and "exiled" metaphysics and music "without difficulty" (Morrell & Thackray 1981: 276).

5. The Great Exhibitions of 1851, 1855, and 1862, but especially the Paris exhibition of 1867, in which "German and French technical achievements were the star attractions" (Engel 1983: 298), demonstrated that Britain was falling behind in science (Burrage 1992: 198; Lawson & Silver 1973: 303). The ongoing imperial expansion, the sharp international trading rivalry (Rothblatt 1983: 132), and the awareness of growing Prussian power between 1866 and 1871 (Haines 1969: 15, 47, 49) led to an increase in state subsidies in science, which made reform in institutions and a professional career in science possible. In France, even before the war of 1870, Victor Duruy, minister of education in the Second Empire, funded research through the École Pratique des Hautes Études after the French government recognized the growing scientific progress of its rivals (H. Paul 1985: 44; Geiger 1980: 358).

6. The few positions available "in the nation's traditional seats of learning, were often monopolized by aristocratic dilettantes . . . [while] scientific charlatans were taking advantage of the unclear status of knowledge" (Alborn 1996: 94). Up until the late nineteenth century, "science in England meant a random and casual familiarity with natural phenomena, rather than a theoretical grasp of a subject acquired through disciplined study" (Berman 1975: 38). For Cardwell it is clear that the denial of state aid during the crucial period 1850–1880 was the main reason that applied science came about later in England than in Germany: "Theories of self-help and of individualism proved, when applied to science and education, incapable of producing the professional scientist" (1957: 187). "By 1900 there was hardly a chemist of any standing in Britain who had not a German PhD—a degree still not obtainable in his own country" (Farrar 1976: 189).

7. C. McClelland asserts that "much more vigorously than in Western Europe, romanticism in Germany transcended the realm of the arts and literature to influence virtually every activity" (1980: 172). This was already apparent before the French Revolution, notably in Jena between 1796 and 1804 with Schiller, Schlegel, and Novalis. Romanticism manifested itself in the love of German customs and tradition (Moser), the celebrated love of the Fatherland (Klopstock), the German Sturm und Drang movement, and, as in France, in an internal critique coming from within the Enlightenment tradition (L. Williams 1973: 3–22). Even more than in England, patriotic opposition to the French aggressor (due to the occupation of Germany) strengthened the appeal of romanticism, as evidenced in the work of Schiller, von Kleist, and Hölderlin (von Engelhardt 1988: 117).

8. The fact that contemporary art historians do not deal with Goethe's "nonliterary work" reflects the continuing split between the two cultures.

9. The romantic study of history and nature is not easily or sharply distinguishable from the speculative *Naturphilosophie* (Rupke 1983: 391; cf. also Culotta 1974: 3). Yet, by lumping various *Naturphilosophen* with Goethe as "the romantic reaction" against Newtonianism, we are not implying that there were no differences between "romantic, speculative and transcendental, scientific and aesthetic directions," as von Engelhardt (1988: 112) points out. They are, rather, "tandem developments" (Snelders 1970: 193). What is important here is that men like Schelling and Hegel considered Goethe as an ally in "opposing the confused night of Newtonian physics" (Stephenson 1995: 29) since "his criticisms of the Enlightenment were grist to the mill of Romantic writers of both literature and science" (Knight 1986: 55). Still, this mode of opposition to Newton was particular to nineteenth-century scientific Germany, since being scornful of Newton in England was regarded as being antiscientific *tout court* (Knight 1976: 167). The "typical reaction of the English physicist" to Goethe's study was indicated by Thomas Young's review of *Zür Farbenlehre* (published in January 1814): he regarded it as "a striking example of the perversion of the human faculties" (Ribe 1985: 315).

10. "Both Hölderlin and Novalis were deeply and dangerously affected by the extreme intensification of this opposition in their own time; by Fichte's solipsism on the one hand, purely mechanistic interpretations of nature on the other. Both eagerly availed themselves of Schelling's doctrine of the identity of world and mind. Especially Novalis 'desired nothing less than to master the whole of modern science and philosophy in order to "poeticize" them, to integrate them into his own ideal of wholeness'" (Hamburger 1957: 72, 75).

11. The same dismissal was made of Bergson about a century later when he attempted a critique of Einstein's relativity theory in his *Durée et simultanéité* (1922).

12. See Durant (1943: 336–50), or Deleuze (1966), who stressses his intuition. Not unsurprisingly, Bergson was also under the indirect influence of German romanticism via Schelling (Vloemans 1966: 17–18; 103–4). For some other exceptions pleading for a reconciliation between science and the arts, see Day (1972/1973: 193).

13. The 1914 lecture "On Scientific Method in Philosophy" was delivered at Oxford (Russell 1954: 7; 95–119). Since Russell "reduced existence to atomic facts, [it followed that for him] values [lay] outside the realm of truth and falsehood" (Lewis 1974: 14).

14. See Whitehead (1947) and Stebbing ([1937] 1944), who discuss in detail the issues related to "Huxley's demon," physical determinism, entropy, human freedom, and responsibility. The German defeat in the First World War also triggered a specific debate on the extent to which natural science was conditioned by the milieu (Forman 1971: 1–115).

15. This is especially clear in the writings of one of the first professors of English literature, F. D. Maurice (1805–1872), who from 1840 on was a professor of English literature at King's College, London. See Maurice (1839).

16. Jules Ferry, one of the most prominent politicians in late nineteenth-century France, stated his goal quite clearly: "The school must shape the soul and the brain of the child to patriotic and national ends" (quoted in Coutel 1996: 971).

17. From the 1880s on, elementary education would be framed by French, history, and geography as the three essential components to identify oneself with the nation (Ponteil 1966: 294). This reform would slowly reach the higher educational system as well.

✳

4

The Social Sciences and Alternative Disciplinary Models

Mauro Di Meglio

One of the most important outcomes of the French Revolution was the emergence of new ideological discourses in the world-system. The very idea of revolution was itself a consequence that implied "a new consciousness of history and a new concept of the social order." In these terms, "it stands at the origin of modern social and historical thinking" (Sewell 1985: 84). From the vantage point of the capitalist world-economy as a whole, the French Revolution marked the moment when the old "feudal" ideological clothes were dropped and the idea that social change was "normal" became widespread. Dismayed by the French uprisings and by the prospect that the popular masses might try to seize state power through similar rebellions elsewhere, the powerful of the world drew the logical inference that "it was only by accepting the normality of change that the world bourgeoisie had a chance of containing it and slowing it down" (Wallerstein 1989b: 43–44).

Once social change was accepted as normal, there was great incentive to try to organize control over such change. The social sciences (as a set of collectively conceived concepts and institutions) emerged in large part as a mode of understanding the nature of these processes in the hope of being able to control them or at least channel them. In these terms, we can consider the emergence, bounding, and institutionalization of the disciplines of the social sciences as one of the world-historical consequences of the ideological transformation brought about by the French Revolution.[1]

The term "social science" (in both the singular and the plural) emerged in the late eighteenth century as one designation sometimes used for what were more frequently at that time called the "political and moral sciences" (see Senn 1958; Iggers 1959; Baker 1969; Shapiro 1984). Until the middle of the nineteenth century, however, the social sciences scarcely existed as academic subjects anywhere. Their academic establishment during the second half of the nineteenth century was slow and uneven and was inscribed within the revival of the role of the universities in the structures of knowledge in nearly all Western countries.

The report of the Gulbenkian Commission (1996) traced the process through which the social sciences became institutionalized as a series of separate disciplines between 1850 and 1945. Each came to occupy a separate domain (defined in terms of both subject matter and proprietary theory and methods) with a specific name. The process of construction was played out primarily in a few Western countries—Great Britain, France, Germany, Italy, and the United States—and took the form of an increasing setting of boundaries delimiting what were considered distinct fields of inquiry.[2] This institutionalization was carried out by establishing in the principal universities chairs, departments, and degrees carrying the names of these disciplines and by creating specialized journals, national and (then) international scholarly associations, and library catalogs arranged according to these disciplinary lines (see Lee 1994). These institutional mechanisms functioned to shape and give legitimacy to the emergent academic departments and demarcate the intellectual fields with which they dealt by providing a means of communication and open discussion for all those prepared to respect the new canons of scholarly debate and to accept the discipline of the boundaries that were implied.

Indeed, the social science disciplines took on their distinctive characters as an outcome of a clash of political agendas that was manifested in a series of methodological debates at the end of the nineteenth century with pressure to line up one way or the other on the by then established split between the two cultures. In the end, each of the final set of disciplines came down, predominantly, either on the nomothetic or the idiographic side. In fact, as the Gulbenkian report concludes, "a very large and diverse set of names of 'subject matters' or 'disciplines' were put forward during the course of the century. However, by the First World War, there was general convergence or consensus around a few specific names, and the other candidates were more or less dropped" (Gulbenkian Commission 1996: 14). Economics, political science, and sociology tended to be more nomothetic. History, anthropology, and Oriental studies tended to be more idiographic.

The adoption of this organizational model of knowledge was not a smooth and straightforward process. Quite the contrary in fact. Value-oriented analyses of social reality (e.g., conservative critiques of industrial civilization and radical appeals for revolutionary change) continued to flow from the humanities. It would be the social sciences, however, that provided the intellectual tools for managing liberal, incremental change, considered to epitomize progress, and subverted arguments for a return to an idyllic past as well as those for rapid, violent transformations. The key injunction was the concept of value-neutrality—imported from the natural sciences—which was a call to undertake social inquiry without recourse to the conflicting value presuppositions on which the analytic traditions in the humanities were founded. Nonetheless, the adaptation of the social science disciplines to the two cultures' division of knowledge was never without debate. As a matter of fact, despite the medium-term resolution of the crisis of the structures of knowledge, the social sciences constituted one of the main loci of struggle between the two cultures.

This process had, furthermore, a national dimension. That is, the common trends stemming from being part of the same world-system were shaped by and in specific national contexts. Thus, in each country social science disciplines took on somewhat different political tasks, and each national history was formed by the interaction of common and national factors. To assert the importance of the national dimension in the emergence and institutionalization of social science disciplines does not simply mean paying homage to the "majestic" role played by nation-states in patronizing social analysis during the last 150 years or so. Nor does it simply mean acknowledging the importance and salience of distinct "national" intellectual traditions, whatever their relative degree of autonomy might have been. The emphasis on the national dimension addresses the importance of the place occupied by particular states in the hierarchical structures of the capitalist world-economy—its axial division of labor and the interstate system.

This national specificity played a key role in the process through which the social sciences were constructed in the period between the mid-nineteenth century and 1945. In fact, this process may most usefully be conceptualized not so much as a linear history but as the history of a continual antagonism between two opposing models: an emerging "hegemonic" and universalistic model, more "disciplined" and with greater emphasis on scientific and positivistic components, and models of resistance (Wallerstein 1977) that stressed a more holistic understanding of the functioning of the system and accorded a particular importance to temporal and spatial awareness. In other words, if, on the whole, the hegemonic model is able to explain the direction followed by this process in the Western world in the period 1850–1945, it is also necessary to recognize the existence of alternative patterns that expressed their distance (or insulated themselves) from the former, not only in terms of strategies of development in the capitalist world-economy, but also in terms of the "national" organization of structures of knowledge.

The construction of the modern academic discipline of history from the early nineteenth century offers a particularly pertinent example. Central to its emergence was the firm belief in its scientific status as a source of authoritative knowledge, which implied a clear-cut division between professional historians and the literary pretensions of amateurs. A scientific orientation meant that historians shared the optimism that methodologically correct research, based on empirical data, made objective knowledge possible. They believed that truth consisted in the correspondence of knowledge to a real and objective world (Iggers 1997). The historiographical revolution associated with Leopold von Ranke that took place in the Germanies emphasized the study of the past *wie es eigentlich gewesen ist,* "as it really happened," and presupposed a strict abstinence from value judgments (Ranke 1973). Modern historians thus joined natural scientists in their struggle against philosophical, that is, metaphysical, speculation. In this sense, history was "in search of science" (Wallerstein 1996).

Yet, at the same time, in Ranke's view, a rigorous historical approach, far from revealing the relativity and hence ethical meaninglessness of existence, reflected a world of meaning and values as they expressed themselves in the historical intentions of human beings and in the values and mores that gave societies cohesion. Ranke's call for studying the past to show how things actually had been referred not to a purely "factual" representation of the past but to the past in its essence, as a deeper reality behind concrete historical phenomena (Manicas 1987: 119). As Ranke puts it, "the historian is merely the organ of the general spirit which speaks through him and takes a real form" (quoted in Iggers 1983: 77).[3]

If this reconciled the tension between the scientific ethos of the historical profession and the political and cultural role of the historian, it also often manifested a commitment to a certain social order, nowhere more apparent than in the legitimation of the nineteenth-century emergence of a unified German state. The emphasis on the self-consciousness of human beings, on their autonomy and unpredictability, and hence on the uniqueness of all events and the impossibility of generalizations, accounts for history's idiographic orientation as well. In fact, the rejection of philosophy by historians was double: first, they asserted the primacy of science, as they conceived it, against medieval modes of knowledge; and second, they refused philosophy, insofar as it entailed the search for general schemas and laws of the social world.[4]

In the realm of social policy, it is hardly a coincidence that in the nineteenth century, *Staatswissenschaften*—the main current of opposition to the orthodox mainstream and its abstraction of purely economic (or political, or sociocultural) processes from the totality of social interactions—originated and flourished in the German-speaking world. On the one hand, the hegemonic position of Great Britain in the nineteenth century accounted for the formulation of the economic principles of free trade and laissez-faire and for the attempt to spread them, with their related assumptions of timelessness and universality regarding economic behavior.[5] On the other hand, the organizational needs of the Germanies, considered from the vantage point of a strategy of development, made such a recipe inapplicable, or in any case ineffective. The German Historical School developed as a countermovement in opposition to classical economics and its timeless and atomistic approach. It denied the existence of natural laws applying to economic activity and emphasized the embedding of the market in the totality of national life.

From this perspective, what was most unacceptable was the exclusion of the state from the analysis of economic and social phenomena, given the pivotal role the German Historical School assigned to it in the process of catching up development—hence the name, Staatswissenschaften. The Historical School's opposition to abstract universalism was linked to German organizational needs in the context of interstate competition. This aspect is best exemplified by the work of Friedrich List. In *The National System of Political Economy* ([1841] 1856), List denounced the economic principles of the classical school, which, he said, reflected the industrial and commercial supremacy of England in the nineteenth century and were inapplicable to the needs of "less advanced" but rising countries, such as nineteenth-century Germany, France, and the United States. He criticized laissez-faire, arguing that behind the slogan of free trade lay the interest of the "strongest nation" to ensure world domination (see Tribe 1995). Instead, he demanded a "national system of political economy" and developed his countertheory of a "confederation of the productive forces" and of economic stages against the individualistic-cosmopolitan orientation of classical economics. He asserted the importance of state protection of agriculture and of infant industries, as well as of government subsidies for shipping and commerce.[6] Central to List's analysis was the belief that the intellectual legacy of Adam Smith was marked by "an artificial divorce of politics from economics" (Tribe 1995: 32). In the long run, List's ideas were adopted by virtually all states that strove to develop an industrial system rivaling that of the leading power (Senghaas 1985, 1991).[7]

In addition to the central role of the state, two other themes were developed in opposition to the dominant model: an emphasis on the use of history, and the rejection of disciplinary partitions. In the Germanies, the emphasis on the historical dimension of social processes as a corrective to any simplistic dogma in economics was very widespread. The dominant German conception of history at this time privileged a basic theoretical conviction in regard to the nature of history (see Iggers 1983).

In the first two-thirds of the nineteenth century, "*Historismus* signified a historical orientation which recognized individuality in its 'concrete tem-poral-spatiality'" (Iggers 1995: 130) rather than in terms of timeless, ab-solutely valid truths that correspond to the rational order dominating throughout the universe (Iggers 1983: 5). In this view, all concrete exist-ence is historical, history is a flux, and all human ideas and values have a historical character—and hence there existed a fundamental difference between the phenomena of nature and those of history.

Consequently, the social and cultural sciences required a quite dif-ferent approach from those of the natural sciences. The nature of a thing was supposed to lie in its history.[8] The theory of historical knowledge characteristic of the German Historical School, as it developed at the universities in the nineteenth century, was founded on these premises. The basic philosophic assumptions upon which the tradition rested were accepted not only by the majority of German historians but also by scholars in other disciplines (see Lindenfeld 1988). As Iggers has pointed out, "The philosophy and methodology of Historicism permeated all the German humanistic and cultural sciences, so that linguistics, philology, economics, art, law, philosophy, and theology became historically orient-ed studies" (1983: 4). In a sense, this generation of thinkers looked to history as a means of comprehending the instabilities of their era (Lin-denfeld 1993: 410; see also Lees 1974: 30, 34ff). They insisted on the developmental character of capitalism, which they said had evolved in a series of stages from other types of economic organization. The very phrase "historical method"—along with others such as "culture" and "society"—functioned as an "embodiment," that is, a "unit-idea," a focal point for a variety of intellectual needs, particularly strong in periods of rapid transition (see Lindenfeld 1988).

Finally, all this was closely connected to the specifically German rejection of the disciplinary partitions that were gaining strength in oth-er western European countries.[9]

> [Staatswissenschaften] covered (in current language) a mixture of eco-nomic history, jurisprudence, sociology, and economics—insisting on the historical specificity of different "states" and making no one of the disciplinary distinctions that were coming into use in Great Britain and France. The very name *Staatswissenschaften* ("sciences of the state") indicated that its proponents were seeking to occupy somewhat the same intellectual space that "political economy" had covered earlier in Great Britain and France, and therefore the same function of providing knowledge that was useful, at least in the longer run, to the states. (Gulbenkian Commission 1996: 18)

If it is true that the German Historical School wrote a history of capitalism quite different from the British narrative, it is also true that it remained entirely within the liberal (or liberal-Marxist) paradigm. In fact, the framework advanced in German scholarship limited itself to introducing changes to the British approach without questioning its basic assumptions. This can be appreciated if we notice that its opposition to the universalism of the hegemonic model was temporally bound. Once again, this temporal limitation can be related to German organizational needs in the struggle for hegemony in the capitalist world-economy. And, once again, List's work shows clearly this aspect.

List did not completely reject the universalistic propositions of classical economics and judged the Smithian cosmopolitan paradigm suitable for a hypothetical stage of joint control of the world market by Western industrialized countries (including, of course, Germany), that is, a goal to pursue for the exploitation of the rest of the world by means of free trade.[10] In fact, in the first decades of the twentieth century German social science finally began to adopt the disciplinary partitions in use in Great Britain and France. The term Staatswissenschaften was gradually replaced by *Sozialwissenschaften* ("social sciences"), and Staatswissenschaftliche chairs at German universities became chairs of sociology, economy, psychology, and later political economy (Strohmayer 1997: 311).[11]

As the German Historical School developed the criteria of objectivity and critical use of archival documents into a "science of history," it also achieved a critical balance between the universality of Ranke's vision and the image of uniqueness and ceaseless change historians actually presented. However, with the rise and expansion of the Prussian state, idealism gave way to the construction of a *Volksgeist* as a foundation for a German nationalism that could sanction unification. The resulting decline of the transcendent element left historicism, as science, open to positivist challenges and charges of relativism. These challenges became more insistent after the middle of the nineteenth century when the "perfect" accord between John Stuart Mill and Auguste Comte became clear. Comte especially appreciated Mill's inclusion of the "indispensable" deductive step in the *System of Logic* (1843), translated into German in 1848. The dilemma was real: history could preserve its objectivity only at the loss of its ethical orientation; otherwise, it would cease to qualify as a producer of systematic knowledge.

Consequently, it was not only in the policy arena that efforts were made to rethink theory and method in social research. The so-called *Methodenstreit*—the debate between the theoretical-analytical approach of the Austrian school led by Carl Menger and the historical-institutionalist approach of the German Historical School led by Gustav von Schmoller—appears as a crucial moment in defining and institutionalizing the social sciences as university disciplines. Formally, the dispute began with Menger's critique of German historical economics, making the case for pure theory based on assumptions about behavior and antecedent conditions (Menger 1883), and Schmoller's review of this book (Schmoller

1883), which argued for principles of economics based on empirical historical data and the inductive method. Of course, these views on the nature and relation of theoretical and practical knowledge in the social sciences were related to more complex disagreements about the nature and scope of economics and its policy implications.

Although both Menger and Schmoller advocated the use of both empirical studies and theory, they disagreed on the emphasis to be placed on each. In particular, Menger's basic argument was that, if both individualizing (historical) and generalizing (theoretical) conceptions of economic phenomena were appropriate, they were so at different levels and were not substitutes one for the other (Tribe 1995: 78). It has been rightly pointed out that "this move effectively created the grounds on which one could legitimize the future existence of distinct disciplines within the social sciences, whose degree of exactitude would depend both on the subject matter and the stage of its theoretical development" (Strohmayer 1997: 301). At the same time, the very occurrence of the Methodenstreit indicated the existence of a community of scholars increasingly conscious of their identity as specialists and professionals, imposing rigorous standards and scientific exclusiveness (Lindenfeld 1997: 256).

No doubt, the Methodenstreit played a major role in the intellectual weakening and institutional destruction of any categorization that cast a skeptical eye on both nomothetic and idiographic claims, making of this antinomy the only possible framework for epistemological debate and reinforcing the compartmentalization of the social sciences, confined to the study of the ever smaller in time and in scope.[12]

If we draw a balance sheet of the experience of German resistance to the liberal pattern of institutionalization of the social sciences, however, it is not easy to call it a defeat, and this at least for two reasons. First, the influence of the German Historical School on the historiography of the twentieth century is indeed remarkable. In France, for example, German historical thought had a great influence on the *Annales* movement, born as a reaction against the premises upon which the institutionalization of the social sciences in the nineteenth century had been based.[13] *Annales*'s historians rejected the conception of history in terms of a movement across a one-dimensional line from the past to the future and offered a very different conception of historical time. They sought to transcend the antinomy between the idiographic and the nomothetic by postulating the multiplicity of social times—rhythms of the *conjoncture* and trends of the *longue durée*. At the same time, they advocated the opening of history to other social sciences.[14] The rebellious thought of the *Annales* school was nourished by a large dose of nationalism, which "provided the underlying passion that sustained its ability to serve as a locus of antisystemic resistance" (Wallerstein 1991b: 194).

The second, and more important, reason to resist the temptation to consider the German experience of resistance a defeat is that to "retreat" on universalistic positions can be a sign of achievement of a privileged position in the world-economy. Ulf Strohmayer (1997) has underscored how the predominance in German territories, from the end of the Napoleonic Wars until the beginning of the twentieth century, of a historicist mode of thinking and the appeal of the Staatswissenschaften owes its existence to the "relative social and economic deficiency" related to the German political structure:

> Not yet a nation-state like France and England or Spain and Sweden, the task of providing concrete and particular means of justifications (or answers) for particular and concrete demands by social elites (or problems) was still secondary in importance to the necessity of creating a general frame of reference from which to depart. [This meant that] unlike in Great Britain and France, the legitimative purpose of what came later to be known as the social sciences was not yet internal to an accepted whole but external to a whole yet to be created. (Strohmayer 1997: 287–88)

At the same time, the very outcome of the Methodenstreit can be understood in terms of the dynamics of state-making.

> Schmoller did not lose the *Methodenstreit* because his position was less scientific than Menger's. . . . Rather, he lost it because his conceptualization of the nature of laws in the social sciences and the organizational structure derived from this consideration did not fit into the needs of an epoch in which the whole had been achieved and the search for salvation in the parts had only begun. In an almost tragic but perfectly understandable manner, given his interest in politics and his position in the *Verein*, he insisted on competing on the wrong turf and tried to become an expert for the whole of a society which still lacked the means to challenge the notion of expertise as a regulative ideal. (Strohmayer 1997: 310)

According to Wolf Lepenies, what distinguished the majority of German scholars from those of countries like France and England was the deep-rooted conviction that bourgeois society was no more than a historical phenomenon and that its analysis could not offer the basis for a natural philosophy of human cohabitation. And, as an example of the difficulties encountered by the diffusion of English and French models of social science in Germany, he observed that sociology was regarded as an "Anglo-French discipline marked by an arrogant claim to knowledge" and therefore as a threat insofar as it was "incapable of doing justice to the special features that characterized Germany" and its history. As a consequence,

> the sociology of Western Europe encountered political resistance in Germany because it did not merely accept the scandalous separation of society and state but actually welcomed it as a precondition of its

scientific justification for existing: it provoked a scientific reaction because it misunderstood the nature of historical phenomena and promoted itself as competing with the science of history. (Lepenies 1988:
235–37)

Nonetheless, to understand the trajectory of German social science
during the nineteenth and the early twentieth centuries, and indeed the
very outcome of the Methodenstreit, we have to take into account not
only the state-making process but also the success of the German strategy of development, which made Germany one of the candidates for
hegemony in the capitalist world-economy. In this perspective, the experience of the United States is a further, and perhaps even more meaningful, case in point.

On January 8, 1918, Woodrow Wilson, in his address to the joint
session of the Congress of the United States, formulated the Fourteen
Points that he regarded as being an essential basis of a postwar peace
settlement. We can reasonably assume that this declaration marked the
new consciousness by the United States—an awareness in part that already existed before the First World War—of its new role on the world
scene, and, consequently, the beginning of the structuring of a new
American internationalism, universalistic in nature (Knock 1992). Ever
since, this universalistic internationalism has occupied a central place in
American ideology and has crucially influenced the social sciences (see
Lentini 1998: ch. 1).

In order to serve the new geopolitical needs of the United States,
American social science had to undergo a process of "de-Germaniza-
tion" (see Manicas 1987). This is not surprising if we remember that the
majority of the nine thousand Americans who studied in Germany between 1829 and 1920 did their studies in the "social sciences" in the last
decades of the nineteenth century.[15] Scholars coming back from Germany were convinced that the German critique of Smithianism was correct.
They therefore rejected laissez-faire and looked to the *Verein für Sozialpolitik* as a model for the social reform of society. Although composed of
republicans and reformers, valuing cooperative self-help and "social ethics," and very far from even the limp socialism of the *Kathedersozialis-
mus*, this first generation of social science scholars was nonetheless
German "in thinking of *Geisteswissenschaft* as a kind of historically oriented unified social science with overlapping, non-discrete, connected
concerns." Moreover, they still believed that policy judgments could be
just as "scientific" as any other judgments (Manicas 1987: 214). The United
States was, as a consequence, characterized by the presence of a distinctly German, historical, and holistic conception of society and history
(Manicas 1991: 48).

The process of the Americanization of the social sciences involved in
the first place the professionalization of social science, that is, the tran-

sition from an ameliorative and associational social inquiry to professionally defined theoretical discourses and scholars' communities based in the universities.[16] The history of the American Social Science Association (ASSA, 1865–1909) offers a propitious vantage point from which to observe not only this transition (see Van Tassel 1984) but also the related process of disciplinarization. In fact, alongside professionalism and decentralized institutions, the United States experienced an increase in the requisite separatism in the social science disciplines (Manicas 1987: 211). The triumph of specialization signaled the transition to a new division of labor in social inquiry, which replaced that introduced by the American Social Science Association.

The ASSA was initially organized in four departments—education; public health; economy, trade, and finance (or social economy); and jurisprudence—on the example of the British National Association for the Promotion of Social Science. This departmental structure conformed, as much as possible, to the existing pattern of specialization within the professional class (Haskell 1977: 104–9). The formation of the various national scholarly associations and academic journals,[17] which gave formal and tangible representation to communities of inquiry rooted in the academic world, signaled the transition to a new division of labor in social analysis, one that paid little attention to the classic professions.

The process of abandoning a German-inspired, comprehensive conception of social-scientific inquiry in favor of pragmatically and narrowly disciplinary discourses, exclusively oriented to the needs of the academic scholar, was to take two generations, beginning with the creation of graduate programs in the social sciences, also modeled on the German example. Manicas argues that the first generation of American social scientists sought to find a solution to two dilemmas: "how to be German professors in a rapidly changing America, and how to be 'scientific' (and thus 'objective') and at the same time have a political impact"; and how to have authority without having some specialized "discipline," that is, the dilemma of social science versus the social sciences (1987: 216). The first generation did not succeed in resolving these dilemmas. The second generation, however, opted for the particular strategy of disciplinary institutionalization and professionalization that rapidly became prevalent in the United States. This process implied the disappearance of all synthetic ambitions for the human sciences across disciplinary lines.[18]

Concomitantly, one after the other, the social sciences came to be dominated by positivistic trends, dedicating themselves to the search for scientific laws based on natural scientific models.[19] In the 1920s, we witness a new structuring of the social sciences, based on a growing scientism, positivism, empiricism, and operationalism and on the progressive abandonment of a historical orientation. Although this process was perhaps not deliberately intended to depoliticize the new social problems and to reduce them to strictly "scientific terms" manageable by so-called experts, it is fair to say that this was one of the consequences of the Americanization of the social sciences (see Manicas 1987: 211–12).

The professional base of scientism was strengthened during the 1920s by the investment of large sums of money by the Carnegie and Rockefeller foundations. Academic social scientists and foundation officials saw the advantages of collaboration: social scientists saw in the philanthropist the fountainhead of resources needed for the construction of a science of society; philanthropists believed that the emerging scientific idiom of social science could provide the sound knowledge necessary for bringing about the reforms they desired.

Funded by Rockefeller money, the formation of the Social Science Research Council (SSRC) in 1923 acted as a catalyst for the focus of social science on scientific method.[20] At the same time, the SSRC institutionalized a pattern of cooperation among the social science disciplines. However, this should not mislead us:

> The ability to join together was itself a sign of confidence on the part of each social science discipline in its own distinctiveness. The research projects funded and devised by the SSRC were not generally interdisciplinary in conceptualization. What brought the disciplines together was their joint concern to promote their own fields (D. Ross 1991: 401).

This trend coincided with the rise of the United States to hegemonic status within the world-system.[21] If up to this point they had made use of a social science supporting their ascent to the core of the capitalist world-economy, they now demanded a kind of knowledge that would be effective in managing the world-system from a dominant position.[22]

However, up to the mid-twentieth century, the diffusion of this model of disciplinary partitions and institutionalization was not undisputed, even in the United States. As a matter of fact, a fully articulated model of "disciplined" social sciences did not become accepted everywhere until after 1945, when the American project completed its expansion and came to prevail in European and Third World countries through the diffusion both of the Parsonian version of social science and of modernization theory.[23]

Many scholars, in the period before 1945, resisted the disciplinary framework (Turner & Turner 1990: 121–23; also Ogburn & Goldenweisser 1927). The encyclopedic projects developed in the first decades of the twentieth century, both in the United States and in Europe, offer a further example of resistance to the diffusion of the liberal paradigm and its disciplinary partitions. In the *Encyclopedia of the Social Sciences,* published in the United States between 1930 and 1935, the central place was occupied by problems, and not by disciplines (Lentini 1998: 14).[24]

Other projects explicitly resisting the liberal model of disciplinarization of the social sciences can be found in Europe. In 1925, Henri Berr founded in France the Centre International de Synthèse. His intention was to foster the growth of a "spontaneous" order of knowledge, replacing the hierarchical orders among disciplines, and oriented toward an open and "unsaturated" encyclopedism, constantly exposed to the destabilizing effects of "human evolution" (Gemelli 1999b: 24).[25] And in the same years, George Sarton, expatriated in the United States during the First World War and founder of the journal *Isis*, argued that the inspiring principle of the encyclopedic synthesis was not the classification of disciplines but their aggregation around the concepts and issues of science; moreover, such aggregation was to be based not on the analysis of concepts but on the crossing of disciplines (Gemelli 1999b: 26; see Sarton 1924).

The methodological and epistemological consequences of these kinds of projects implied the drawing of "cross-disciplinary" conceptual maps and the reversal of the very logic of interdisciplinarity, interpreted as the mechanical transfer of techniques from one disciplinary field to the other, rather than as the focusing on issues around which to aggregate, each time, the disciplinary dialogue. However, as Gemelli has pointed out, Berr and Sarton did not overcome, in their work on "interscience," the separation between the two cultures. In fact, it was Lucien Febvre, starting his project of the *Encyclopédie Française* in 1932 and operating in an even more radical perspective, who came closer to this model of knowledge, purposely overcoming the idea of a hierarchical organization of disciplines (Gemelli 1999b: 27).

What conclusions can be drawn about the emergence of the social science disciplines between the mid-nineteenth century and 1945? First of all, there was an overall trend in the direction of an increasing differentiation and institutionalization of a few distinct fields of inquiry. However, this tendency to impose the dominance of a sectorializing (and universalizing, whether nomothetic or idiographic) thought was not uncontested. The problems inherent in the process of "disciplinarization" of social analysis and the pervasiveness of various movements of resistance suggest that the history of social science and its disciplinary structuring is not so much a linear process of progressive and homogeneous organization as the history of the struggle between a tendency to impose the dominance of the epistemological principles of universalizing-sectorializing thought, and a tendency to transcend the opposition between the universalistic and particularistic approaches to knowledge and the distinction between the two cultures. Furthermore, this history is linked to the dynamics of the capitalist world-economy, and particularly to the national strategies of development in the context of interstate competition.

Notes

1. This is not the usual way of explaining the emergence of the social sciences. The usual history tells the story as part of the linear process of the Enlightenment, an

internally driven progress of reason. In essence, this mode of argument writes the history of a discipline backward, taking as definitive the present presumed consensus and reconstituting the past as a teleology leading up to and fully manifested in it (see Collini, Winch, & Burrow 1983: 3–7).

2. Social science was institutionalized, in the course of the nineteenth century through the differentiation of the existing European university structure, which, at the end of the eighteenth century, was still largely organized in the traditional four faculties of theology, medicine, law, and philosophy. In this process, "[t]he faculty of theology became minor, sometimes disappearing completely, or being replaced by a mere department of religious studies within the faculty of philosophy. The faculty of medicine conserved its role as the center of training in a specific professional domain, now entirely defined as applied scientific knowledge. It was primarily within the faculty of philosophy (and to a far lesser degree within the faculty of law) that the modern structures of knowledge were to be built. It was into this faculty (which remained structurally unified in many universities, but was subdivided in others) that the practitioners of both the arts and the natural sciences would enter and build their multiple autonomous disciplinary structures" (Gulbenkian Commission 1996: 6–7).

3. Iggers observes: "The 'impartial' way of looking at things for which Ranke argued in fact revealed the ethical character of social institutions as they had developed historically. Although replacing Hegel's philosophic approach by a historical one, Ranke agreed with Hegel that the existing political states, insofar as they were the results of historical growth, constituted 'moral energies,' 'thoughts of God.' . . . The 'impartial' approach to the past . . . thus in fact for Ranke revealed the existing order as God has willed it" (1997: 26). On the apparently contradictory coexistence in Ranke's thought of a transcendental faith and a passion for particular truths, that is, on the coexistence of universality and individuality intended as antinomies of life, see Krieger (1975).

4. As Wallerstein has pointed out: "Historians were haunted by their image of philosophy, and of what was called the philosophy of history. They had rebelled against philosophy, which was seen as deductive, and therefore speculative, and therefore fictional or magical. In their struggle to liberate themselves from the social pressures of hagiography, they insisted on being empirical, on locating 'sources' of real 'events.' To be nomothetic was to 'theorize' and therefore to 'speculate.' It was to be 'subjective,' and therefore to go beyond what was knowable or, worse, to recount reality incorrectly and prejudicially" (1996: 12–13).

5. It must, however, be remembered that in Great Britain, with the waning of hegemony between the 1870s and the 1930s, English historical economists challenged the theory, policy recommendations, and academic dominance of classical and neo-classical economics. They succeeded in establishing economic history as a separate and academically recognized discipline. Their work was, among other things, a response to the growing political and economic competition of the United States and Germany (see Koot 1980, 1987).

6. List's critique of "cosmopolitan economy" drew from his acquaintance with U.S. economic debates of the late 1820s (see Notz 1926). The basic elements of his "post-Smithian" political economy—his conceptions of economic protection based upon the "infant industries" thesis and that of national economic growth within an international order—were shaped in the period 1825–1828 in the United States and subsequently "imported" by Germany (see Tribe 1995). As Seligman argues, "Hamilton may well be called the spiritual father of Friedrich List" (1925: 134).

7. The Verein für Sozialpolitik, founded in 1873, was a group of scholars and public figures whose chief reform proposal was state intervention in the economy.

"We are convinced that the unrestricted play of contrary and unequally strong private interests does not guarantee the common welfare, that the demands of the common interest and of humanity must be safeguarded in economic affairs, and that the well-considered interference of the state has to be called upon early in order to protect the legitimate interests of all. We do not regard state welfare as an emergency measure or as an unavoidable evil, but as the fulfillment of one of the highest tasks of our time and nation. In the serious execution of this task the egotism of individuals and the narrow interest of classes will be subordinated to the lasting and higher destiny of the whole" (in *Jahrbücher für Nationalökonomie und Statistik*, 21 [1873], 123, quoted in Herbst [1965] 1972: 144–45). On the history of the Verein für Sozialpolitik, see Roversi 1984.

8. Mandelbaum has synthesized this approach as follows: "Historicism is the belief that an adequate understanding of the nature of any phenomenon and an adequate assessment of its value are to be gained through considering it in terms of the place it occupied and the role which it played within a process of development" (Mandelbaum quoted in Ankersmit 1995: 143–44).

9. E.g., French social science developed as a tripartite structure composed of "political science," "economic science," and "human science" (Heilbron 1991).

10. List's problem was the achievement of a "fair" distribution among industrialized countries of the advantages of the world market. He talked about a "civilizing effort" and was persuaded of the utility of free trade, once Germany could participate in the partitioning of the spoils. The imperialist program of German capitalism was thus grounded in social analysis.

11. "Much of the interchange between economics, philosophy, history, and biology took place under the umbrella term of sociology. While we are inclined to think of sociology as a newly emerging specialty within the field of the social sciences, this makes more sense for the 1920s than for the Wilhelminian period. With the possible exception of Tönnies, the so-called founding fathers of German sociology (Simmel, Max Weber, and Sombart) were much less goal-directed in laying the foundations for such a specialty than is usually thought. Rather than being self-conscious advocates of a particular discipline, they were drawn to a set of substantive issues and problems (such as the nature of capitalism) which were larger than could be handled by any single approach. As they worked through these issues, they gradually became aware of a greater need for clearer concepts and methods; the term sociology expressed the need as well as the initial tentative results. . . . The founding of the German Sociological Association in 1909 admittedly marked a turning point in the institutionalization of the discipline, but the variety of positions and approaches aired in its meetings revealed how gradual the process continued to be" (Lindenfeld 1997: 296; see also Wagner 1991). Wagner argues not only that sociology was not institutionalized in Europe in its "classical era" but also that no common understanding on what a science of society should be was achieved, and although standards of sociological work were developed and proposed they could not be enforced among those who considered themselves sociologists.

12. See Wallerstein (1997b). However, as Lepenies has observed, the divorce between the idiographic and the nomothetic epistemologies, or between the attitude of literary men and that of scientists, "did not proceed in a straight, undeviating line, but was characterized rather by the difference in the pace at which it took place in the different disciplines: it did not encompass every discipline, and those it did encompass it affected with a different degree of intensity. National characteristics played an instructive role" (1988: 3).

13. *Annales d'histoire économique et sociale*, the journal founded in 1929 by Lucien Febvre and Marc Bloch that gave the movement its name, "was a direct (and deliberate)

translation of the title of the German review which incarnated the Schmoller school of 'institutional' history, the *Vierteljahrschrift für Sozial- und Wirtschaftsgeschichte"* (Wallerstein 1993: 17).

14. Commenting on the choice for the cover of the *Annales* of the word social—"one of those adjectives which one uses to say so many things over the course of time that in the end it says practically nothing"—Lucien Febvre admitted: "We were in agreement on the idea that a word as vague as 'social' seemed specifically to have been invented and decreed by historical Providence to serve as the emblem of a journal that claimed to be surrounded by no walls, but rather to shine widely, freely, even indiscreetly, on all the neighboring gardens, a spirit, its spirit: I mean a spirit of free criticism and of initiative in all directions" (Febvre 1953: 19–20).

15. See Herbst (1972). "America's first professional social scientists were Americans educated in Germany, and the founding of the institutionalized social science in America was in its first stage fully under the dominance of scholars who had been trained in Germany" (Manicas 1987: 213).

16. According to Haskell, the point of departure for the professionalization of social science was a pervasive mood of doubt and uncertainty, mainly rooted in the intellectual quicksand of an increasingly interdependent social universe and the resultant changes in the conditions of adequate explanation (1977: 47). Dorothy Ross has, on the other hand, argued that the professionalization of American social science should be understood not only as a dimension of this search for authority but also as a product of the peculiar structure of American society, which had an expanding, decentralized university system based on capitalist and middle-class support (1991: 160–61; 1979).

17. The American Historical Association (1884) and its *Papers* (1886); the American Economic Association (1892) and its *Publications* (1886); the American Psychological Association (1892); the American Academy of Political and Social Science (1889); the American Anthropological Association (1902); the American Political Science Association (1903) with its *Proceedings*; the *American Journal of Sociology* (1895); and the American Sociological Society (1905). *Political Science Quarterly* had been established by Columbia University in 1886, and Harvard had launched the *Quarterly Journal of Economics* that same year.

18. See K. Kim (1997: 428). "The second generation, students of that German-trained first generation, decided to be American professors in America. They marked out their territories, defined social science in positivist and ahistorical terms, and adopted an unabashed technocratic stance. This is the difference between the younger Albion Small and the older Albion Small, between Boas and Lowie, between Burgess and Merriam, between Ely and Irving Fisher, between William James and J. B. Watson. After the catastrophe of World War I, it seemed clear to every one that they had made the right choice. Thenceforth, Americans would set the style in "social science"'" (Manicas 1987: 216).

19. "Neoclassical economists first found a paradigm, modeled on physical science and presumably rooted in the necessities of nature and human nature, that exempted American experience from history, at least any history other than the track of perpetual liberal progress that could be extrapolated from the paradigm itself. Institutional economists, most sociologists, and some political scientists, more deeply strained by the rapidity of change and the insecurity of American ideals, sought a different kind of science, an empirical science of the changing liberal world that would allow them technological control. The anxiety to control the careening new world on the one hand, and the narrowed focus and comfortable opportunities of professionalism on the other, turned that scientific impulse toward scientism. Social scientists began to

construct a naturalistic social science as an end in itself and, under the influence of instrumental positivism, erected positivistic scientific method into the chief standard of inquiry. . . . Only the disciplinary tradition of historico-politics and political science had nurtured sufficient respect for and knowledge of history and philosophy to generate a considerable resistance, and there the resistance was vitiated by traditionalist complacencies and fears" (D. Ross 1991: 467–68, and ch. 10).

20. The SSRC-sponsored volume *Methods in Social Science* (Rice 1931) conveyed the message that, up to that point, social science had been insufficiently scientific.

21. "[T]he blind pursuit of rigorous method, precise and exact objectivity, 'universal' laws regardless of time and space, and technocratic control and domination, which eventually led to a reified and mechanistic world-view, have thrived on the basis of hegemonic world power. It is very interesting to find that this increasing nomothetic trend roughly coincided with the shift of the United States' status within the capitalist world-system. As the U.S. became the world hegemonic power, the academic sciences have increasingly reinforced their nomothetic character until both were seriously challenged in the 1970s. These nomothetic sciences including history shared their fate with the demise of the United States' world hegemonic power. In other words, the powerful nomothetic path both in the social sciences and in history reflected the needs and demands of U.S. hegemony in the world-system" (K. Kim 1997: 459).

22. In a sense, the military victory in the First World War was crucial to the triumph of positivism: "For Anglo-Americans, the defeat of Germany represented, as well, the defeat of 'metaphysical,' 'statist,' historical and holistic German social science. Long suspicious of it in any case, the war proved to them that older British and French empirical philosophies, continuously represented in the 'old' political economy and in British utilitarian theories of government, had been right all along" (Manicas 1991: 51).

23. Modernization theory was the response of U.S. social science to the post-1945 geopolitical context and the demands of Third World countries for development. These analyses provided a model of development based on the belief in the idea of progress, rationality, and industrialization, on the basis of an Eurocentric correspondence between modernization and Westernization. The outcome was an interdisciplinary school, in which, however, the various contributions honored the liberal division of labor between economy, polity, and society (see Weiner 1966).

24. "The entries [of the *Encyclopedia of the Social Sciences*] are always inspired by an historical perspective, aimed to highlight the changeable nature of social phenomena, although this seems to be a common characteristic of all encyclopedias of that period. Moving from the premise that there exists a fundamental interdependence among social sciences, the encyclopedia is not so much an exhaustive portrayal of all subjects, as the plan of connections among the various disciplines, largely understood as organizational fields, achieved through a planned cooperation among all the social scientists of the world" (Lentini 1998: 261–62).

Compare this with the *International Encyclopedia of the Social Sciences*, published in 1968, in which there is a strong emphasis on the "disciplinarization" of the social sciences, as well as greater scientism, a renunciation of history and humanistic components, a claim to universality, and an emphasis on quantitative methods and techniques. However, there is no entry for the term "discipline." As Giuliana Gemelli has pointed out, "if [the *Encyclopedia of the Social Sciences*] was based on the principle of differentiation between the method of the social sciences and that of the physical-natural sciences, the *International*, through the functionalist channel, modeled itself on a fundamental adhesion to the imperatives of quantification and formalization, considered as the source of the superiority of the physical-natural sciences in comparison with the social-historical sciences" (1999a: 162). And Ralph Ross has observed that "[t]he *International Encyclopedia* is the fullest statement of what a (perhaps value-free) social science, modeled on physical science, has accomplished in our time" (1976: 942).

25. Berr's very idea of synthesis, of unification of knowledge, "answers the need, at once conscious and obscure, of breaking all the irritating divisions and achieving new and variegated combinations" (Berr 1899: 412, quoted in Gemelli 1999b: 24).

5

The Ambivalent Role of Psychology and Psychoanalysis

Mark Frezzo

In the modern world, psychology as a discipline has remained divided despite repeated attempts by its practitioners to unify it. Throughout its history, "psychology has been pulled in two directions, one analyzing into smaller elements, the other integrating into larger systems" (Miller 1992: 40). E. G. Boring ([1929] 1950: 737–45) once called them the biotropic and sociotropic poles. Advocates of the "biotropic" approach have tended to reduce psychology to neurophysiology. Advocates of the "sociotropic" approach have tended to reduce psychology to sociology. Consequently, despite its success in extricating itself from the faculty of philosophy and establishing an institutional niche for itself, psychology has never succeeded in creating a common framework for the study of the "psyche."[1]

There is little doubt that the tension between the biotropic and sociotropic poles reflects a more fundamental problem: the difficulty of capturing the relationship between the "brain" (conceptualized as a "natural" entity) and the "mind" (conceptualized as a "social" entity). It is not surprising, therefore, that psychology has served as a battleground for the two cultures: whereas biotropic psychologists have deployed experimental techniques to elucidate the functions of the brain, sociotropic psychologists have invoked clinical techniques to explicate the properties of the mind. In short, owing to the intractability of the brain-mind problem, the biotropic and sociotropic approaches have remained antagonistic (if not incommensurable). Pribram has argued that "there are many experimental findings that relate brain, behavior, and experience. Psychophysics, psychophysiology, and neuropsychology abound with illustrations of the relationship between brain and mind, provided one is willing to infer mental constructs from instrumental behavior and verbal reports of experience" (1992: 710). However, many psychologists have questioned the scientificity of such inferences (not least because they entail reference to subjective experience): "radical behaviorists [have] eschewed this readily available solution," opting "to become materialist, physicalist, and 'thoroughly scientific'" (Pribram 1992: 700).[2] It is clear, therefore, that the brain-mind problem continues to plague the discipline.

The "Constitutive Problem" of Psychology

It is instructive that most historians of psychology have grappled with the constitutive problem of the discipline.[3] Despite the intentions of its putative founders—Wilhelm Wundt (1832–1920), William James (1842–1910), and Sigmund Freud (1856–1939)—psychology has never been a unified discipline.[4] On the contrary, owing to widespread disagreement about the scope of the discipline, psychological research—understood as the systematic investigation of immediate experience, consciousness, unconscious processes, and behavior—has always been divided among the faculties of philosophy, science, and medicine.[5] Consequently, such eminent historians of psychology as E. G. Boring and Sigmund Koch have found themselves plagued by the intractable question of the criteria by which one can determine what counts as "psychological knowledge."[6]

It is noteworthy that E. G. Boring's *History of Experimental Psychology* ([1929] 1950), which remains the authoritative text on the emergence of the discipline, underwent considerable revision between its first and second editions (not least because the discipline itself had undergone significant changes in the intervening period). In the preface to the first edition, Boring attempted to demarcate the boundaries of experimental psychology: "Naturally, the words 'experimental psychology' must mean, in my title, what they meant to Wundt and what they meant to nearly all psychologists for fifty or sixty years—that is to say, the psychology of the generalized, human, normal, adult mind as revealed in the psychological laboratory" ([1929] 1950: x).[7] Doubtless, Boring's initial definition of experimental psychology, which emphasized laboratory research on "normal" adults (while accepting data derived from animal studies), reflected the consensus of the period. It included the contributions of the major knowledge movements in academic psychology: psychophysics (1860s–1880s), structuralism (1870s–1920s), functionalism (1870s–1920s), and behaviorism (1910s–1940s).[8] However, in accordance with the prevailing bias for the laboratory as the locus of scientific discovery, Boring's initial definition, colored as it was by the influence of structuralism, excluded the contributions of the psychoanalytic movement (the years 1900–1940s). Boring believed that there were intrinsic limits to the scientificity of psychoanalysis. He thought that such concepts as the superego, ego, and id, though potentially useful as heuristic devices, could be rendered operational only with considerable difficulty. Furthermore, Boring was not convinced that key hypotheses of psychoanalysis—for example, repression, condensation, and transference—had shown themselves to be amenable to experimental verification.

In the second edition in 1950, Boring expanded the definition of experimental psychology to embrace recent developments in dynamic psychology—"a movement for those psychologists who, disappointed in the description of consciousness and behavior, [sought] a more satisfying psychology of . . . 'human nature'" ([1929] 1950: 692). Interestingly enough, Boring classified psychoanalysis, which he had hitherto excluded from consideration, as a form of dynamic psychology, comparing it to the hormic psychology of William McDougall, the cognitive behaviorism of Edwin B. Holt, and the purposive behaviorism of Edward Chace Tolman.[9] Indeed, Boring went so far as to credit Freud with setting the agenda for dynamic psychology: "It was Freud who placed the dynamic conception of psychology where psychologists could see it and take it. They took it, slowly and with hesitation, accepting some basic principles while rejecting many of the trimmings" ([1929] 1950: 707). Thus, Boring came to include psychoanalysis for two reasons. First, despite its marginal position vis-à-vis the university and medical establishments, psychoanalysis had exerted considerable influence on the practices of academic psychologists and psychiatrists. In addition, psychoanalytic concepts had proliferated across popular culture (literature, film, music, and public discourse), leading Boring to believe that Freudianism reflected the zeitgeist. Second, the early attempts to operationalize psychoanalytic concepts, and hence to assimilate the insights of psychoanalysis into behaviorism, harbored the promise of psychology's unification.

Boring was fully aware of the psychoanalytic movement's indeterminate position among the faculties of philosophy, science, and medicine: "As the movement spread beyond [Freud's] control with the defection of Adler, Jung, and finally even Rank, it became an in-between field, a science and a means of therapy, which was accepted by neither the academic psychologists nor the medical profession" (Boring [1929] 1950: 707). Nevertheless, Boring came to the conclusion that the psychoanalytic movement had influenced the trajectory of the entire discipline, not only by providing experimental psychology with an able sparring partner, but also by providing a wealth of concepts (e.g., cathexis and anticathexis) awaiting formalization. Consequently, the second edition of Boring's book displayed a more balanced view of psychoanalysis and, by extension, a more profound comprehension of the constitutive problem of psychology. On the one hand, Boring questioned the scientificity of actually existing psychoanalysis:

> What psychoanalysis will have done for experimental psychology, when the doing has got into history, is another matter. We can say, without any lack of appreciation for what has been accomplished, that psychoanalysis has been prescientific. It has lacked experiments, having developed no technique for control. In the refinement of description without control, it is impossible to distinguish semantic specification from empirical fact. ([1929] 1950: 793)

On the other hand, Boring celebrated the potential of psychoanalytic research: "Psychoanalysis had provided hypotheses galore, and, since operational definitions of its terms are possible, many of its hypotheses can have their consequences tested out by the hypothetico-deductive method. Long ago experimental psychology took the will out of its texts, leaving the space for connotation blank. Now, thanks to Freud, it has got motivation back in" ([1929] 1950: 713).

Boring thus posed the constitutive problem of the discipline in terms of the tension between behaviorism (understood as the emblem of the biotropic pole of psychology) and psychoanalysis (understood as the emblem of the sociotropic pole of psychology). Boring elucidated not only the common "origin" of behaviorism and psychoanalysis (i.e., nineteenth-century psychophysics),[10] but also the considerable cross-pollination between the two prevailing tendencies in psychological research. Accordingly, Boring's narrative stressed the various attempts either to incorporate psychoanalytic insights into behavior theory or to verify psychoanalytic propositions through laboratory investigation.

It is worth noting, however, that there was a significant lacuna in Boring's analysis: the relationship between psychology and the other academic disciplines (especially the social sciences). More precisely, owing to his emphasis on the conflicts within psychology, Boring paid scant attention to the implications of the sciences-humanities antinomy for the university system. Why were most behaviorists operating under the auspices of the faculty of science? Why were most psychoanalysts working beyond the confines of the university system? Why had the attempts to reconcile behaviorism and psychoanalysis failed? To what extent had the rise of positivism, as evidenced in the pervasive influence of the Vienna Circle, exacerbated the tension within psychological research?

It was only after the publication of C. P. Snow's *Two Cultures and the Scientific Revolution* (1959) and Thomas Kuhn's *Structure of Scientific Rev-*

olutions (1962) that philosophers, historians, and sociologists of science began to reflect upon the ambiguous disciplinary status of psychology. It is clear, however, that the constitutive problem of psychology, which has its roots in the nineteenth-century German university system, can be understood only in light of the long-standing tension between the *Naturwissenschaften* and the *Geisteswissenschaften*.

As Mitchell Ash has noted, "the methods of philosophy and of classical philology, not those of the natural sciences, were the original models for the German concept of *Wissenschaft,* and thus for the professionalization of German academic teaching and research" (1980: 259). Thus, in effect, the debate between the Naturwissenschaften and the Geisteswissenschaften, which also played itself out on the terrain of psychology, involved different conceptions of "scientificity." Dilthey, who felt that psychology should be institutionalized as a Geisteswissenschaft, "sharply criticized what he called the 'dominant' psychology, which would 'explain the constitution of the mental world' mechanistically, according to hypotheses about its components, forces, and laws, 'in the same way as physics and chemistry explain the physical world'" (Ash 1980: 270). Moreover, Dilthey sketched an alternative to the emerging scientism: "[Psychology] combines awareness and observation of ourselves, understanding of other people, comparative procedure, experiment, and the study of analogous phenomena. It seeks entry into mental life from many gates" (Ash 1980: 270). Doubtless, Dilthey's distinction between "explanation" (the goal of the Naturwissenschaften) and "understanding" (the goal of the Geisteswissenschaften) betrayed an acute understanding of the constitutive problem of psychology: "We explain nature; we understand the life of the mind" (quoted in Ash 1980: 270). Nonetheless, Dilthey represented a minority in rejecting the idea of psychology as a natural science.

Writing in the aftermath of C. P. Snow's lecture of 1959, Koch lamented the widespread refusal of academic psychologists to acknowledge the "gulf of mutual incomprehension" between the sciences and the humanities: "There has been, at least until very recently, an ever widening estrangement between the scientific makers of human science and the humanistic explorers of the content of man" (1961: 630).[11] In fact, Koch insisted that academic psychology had no choice but to enter the debate on the two cultures: "If psychology is to live up to the purview of its very definition, then it must be that science whose problems lie closest to those of the humanities; indeed, it must be that area in which the problems of the sciences, as traditionally conceived, and the humanities intersect" (1961: 629).

It remained for Koch—in his dual capacity as historian of psychology and philosopher of science—to explain why psychology had failed to confront the sciences–humanities antinomy. Ever since its emergence as an independent field, psychology has been far more concerned with being a science than with a courageous and self-determining confronta‐ tion with its historically constituted subject matter. Its history has been largely a matter of emulating the methods, forms, and symbols of the established sciences, especially physics (Koch 1961: 629–30). Owing to its protracted struggle to establish autonomous institutional structures (e.g., academic departments, laboratories, and research centers), psychol‐ ogy had come to exaggerate its scientificity.[12]

More precisely, even though social scientists, especially sociologists, political scientists, and economists, borrowed heavily from psychologi‐ cal research, experimentalists were not content to make psychology merely the foundation of the social sciences.[13] On the contrary, they wished not only to codify psychology's independence from philosophy, but also to institutionalize the discipline as a natural science. It was widely believed that this could be achieved through the expulsion of such philosophical terms as mind, experience, and even consciousness from the lexicon, and the application of scientific methods to the subject matter of psy‐ chology. Needless to say, behaviorism[14] represented the culmination of this tendency:

> Psychology as the behaviorist views it is a purely objective experimen‐ tal branch of natural science. Its theoretical goal is the prediction and control of behavior. Introspection forms no essential part of its meth‐ ods, nor is the scientific value of its data dependent upon the readiness with which they lend themselves to interpretation in terms of con‐ sciousness. The behaviorist, in his effort to get a unitary scheme of animal response, recognizes no dividing line between man and brute. The behavior of man, with all its refinement and complexity, forms only a part of the behaviorist's total scheme of investigation. (John Broadus Watson, quoted in O'Donnell, 1985: 13)

Hence the behaviorist revolution, which appeared amid the "positiviza‐ tion" of psychology, involved not only the abandonment of introspec‐ tion in favor of more objective experimental techniques, but also the repudiation of consciousness and other "unobservable entities" in favor of behavior. As a consequence of the pervasive influence of Watson and other behaviorists, scientific psychology proscribed all metaphysical spec‐ ulation, thereby actualizing Helmholtz's vision of a "psychology with‐ out a soul."

Psychoanalysis and Experimental Psychology: Failed Rapprochement

Freud's history of the psychoanalytic movement, written in 1914, and Jones's biography of Freud (1953–1957) both offered insight into the relationship between the prevailing tendencies in psychological research. They both emphasized the period 1908–1913, which witnessed the advent of the psychoanalytic movement, the famous conference at Clark University, and the ruptures with Alfred Adler and Carl Jung. Encouraged by the rapid proliferation of psychoanalytic ideas, Freud remained optimistic about the prospects for a dialogue between psychoanalysis and scientific psychology.[15] It was only with Jung's departure from the International Psychoanalytic Association and John Broadus Watson's famous lectures at Columbia University, both in 1913, that Freud abandoned the hope that psychology could become a unified discipline.

The year 1908, which witnessed the emergence of a "Freud Group" at the Burghölzli Mental Hospital in Zurich, represented a significant breakthrough in the history of the psychoanalytic movement: "It appeared that psychoanalysis had unobtrusively awakened interest and gained friends, and that there were even some scientific workers who were ready to acknowledge it" (Freud 1957: 14:26). Most prominent among the scientific workers interested in the nascent discipline of psychoanalysis were Eugen Bleuler, Max Eitingon, and Jung. Reflecting upon the history of the psychoanalytic movement, Freud would later commend these figures for introducing psychoanalytic techniques into medical practice and thereby laying the groundwork for significant advances in the diagnosis and treatment of mental illness:

> The [Zurich school] showed that light could be shed on a number of purely psychiatric cases by adducing the same processes as have been recognized through psychoanalysis to obtain in dreams and neuroses (Freudian mechanisms); and Jung [1907] successfully applied the analytic method of interpretation to the most alien and obscure phenomena of dementia praecox [schizophrenia], so that their sources in the life history and interests of the patient came clearly to light. (Freud 1957: 14:28)

Freud also lauded the Zurich school for attempting to bridge the gap between psychoanalysis (represented by Freud and his associates in Vienna) and experimental psychology (represented by Wundt and his associates in Leipzig): "By this means it had become possible to arrive at rapid experimental confirmation of psychoanalytic observation and to demonstrate directly to students certain connections which an analyst would only have been able to tell them about" (Freud 1957: 14:28). It is clear, therefore, that Freud hoped to participate in the construction of a unified science of psychology. Firmly entrenched in the faculty of medicine, and hence autonomous from the faculties of philosophy and science, psychiatry presented itself as the most promising terrain for the convergence of psychoanalysis and experimental psychology.

It was under the auspices of the Zurich school that Jung invited practitioners and fellow travelers of psychoanalysis—primarily from Switzerland, Austria, and Germany—to found an international organization. Thus, the First Congress of the International Psychoanalytic

Association, which was held in Salzburg on April 26, 1908, provided the occasion not only for the unification of the two psychoanalytic tendencies but also for "the first public recognition of Freud's work" (Jones 1955: 40). In the aftermath of the congress, Bleuler, Freud, and Jung began to publish the *Jahrbuch für psychoanalytische und psychopathologische Forschungen*, with a view to making psychoanalytic thought more accessible to medical doctors (especially psychiatrists and neurologists), university professors (especially experimental psychologists and physiologists), and educated laymen in the German-speaking countries. Unbeknownst to Freud and Jung, however, the psychoanalytic movement was beginning to pique the interest of mental scientists in the United States.

In 1909, Freud and Jung lectured at Clark University, and the history of the psychoanalytic movement reached a turning point: the initiation of a dialogue between psychoanalysis and American experimentalism. As Saul Rosenzweig has noted, G. Stanley Hall[16] hoped that the visit of Freud and Jung would contribute to the reconciliation of the two tendencies and thereby create the possibility of exploring the "totality of human behavior" (Rosenzweig 1992a: 155).[17] Hall believed that psychoanalytic theories—for instance, the theory of repression—could be verified according to experimental procedures.[18] Moreover, Hall subscribed to the view that psychoanalysis would begin "where introspection [left] off and use its findings as its data or point of departure and not regard the [psychic structures] of the post-Wundtians as ultimate" (G. Hall 1923: 412). Thus, in describing the circumstances under which psychoanalysis was officially introduced to the American academy, Freud emphasized Hall's receptiveness to the ideas of the Vienna and Zurich schools:

> In 1909 G. Stanley Hall invited Jung and me to America to go to Clark University, Worcester, Mass., of which he was President, and to spend a week giving lectures (in German) at the celebration of that body's foundation. Hall was justly esteemed as a psychologist and educationalist, and had introduced psycho-analysis into his courses several years earlier; there was a touch of the "king-maker" about him, a pleasure in setting up authorities and then deposing them. (Freud 1959: 20:51)

Owing to favorable encounters with mental scientists at the Clark conference, Freud was sanguine about the possibility of a rapprochement between psychoanalysis and experimental psychology. Indeed, he was particularly encouraged by his encounter with James J. Putnam, a professor of neuropathology at Harvard University, "who in spite of his age was an enthusiastic supporter of psychoanalysis and threw the whole weight of a personality that was universally respected into the defense of the cultural value of analysis and the purity of its aims" (Freud 1959: 20:51).[19] The Third Congress of the International Psychoanalytic Association, held at Weimar on September 21–22, 1911, was the occasion of Adler's secession. Whereas Freud had anticipated that Adler would "apply himself to the connections of psychoanalysis with psychology and to discovering the biological foundations of instinctual processes," Adler elaborated an "individual psychology," thereby leaving the confines of psychoanalysis proper (Freud 1959, 20:61). Hence Freud came to reproach Adler not only for denying the importance of sexuality but also for "[concentrating] on the sociological aspects of consciousness rather than on the repressed unconscious" (Jones 1955: 134).

The Fourth Congress of the International Psychoanalytic Association, held at Munich on September 7, 1913, saw Jung's secession. In the months leading up to the congress, Jung had been downplaying the importance of sexuality, arguing, for example, that incest wishes should not be taken literally. In response, Freud "pointed out to Jung that his conception of the incest complex as something artificial bore a certain resemblance to Adler's view that it was 'arranged' internally to cover impulses of a different nature" (Jones 1955: 147).

Arguably, the ruptures with Adler and Jung, though partly personal in nature, contributed to Freud's disenchantment with the intellectual establishments of Vienna and Zurich. As Jones notes, in response to the contention that there existed three schools of psychoanalysis, Freud argued that "he had the right to know what psychoanalysis was, and what were its characteristic methods and theories that distinguished it from other branches of psychology" (Jones 1955: 151). Hence Freud found himself presiding over the trajectory of psychoanalysis in conspicuously unscientific fashion.

There is little doubt, therefore, that Freud's position on the falsifiability of psychoanalytic propositions, and by extension on the scientificity of psychoanalysis, changed in the course of his career. Whereas in 1912 Freud expressed enthusiasm about Karl Schötter's "experimental confirmation" of the dream theory, in 1934 Freud questioned the utility of Rosenzweig's "validation" of the theory of repression: "I cannot put much value on such confirmation because the abundance of reliable observations on which these propositions rest makes them independent of experimental verification" (Rosenzweig 1992b: 171–73). It follows that the mature Freud believed that psychoanalytic propositions were not amenable to proof under experimental conditions.

Thus it can be argued that:

> The relationship between psychoanalysis and the rest of psychology was a bit stormy from the beginning. The structural psychology of consciousness that dominated the last part of the nineteenth century was not prepared for the unconscious or for a motivational theory of the thought processes. When Freud added the subject matter of sexual development to the cognitive dynamics of dream theory, there seemed no possible bridge between the two kinds of psychology. (Sears 1992: 208)

Structuralism's inability to respond to the demands of psychoanalysis contributed to its demise. Sears argues that it was only with the emergence of behavior-oriented psychology that psychoanalysis found a worthy interlocutor across the divide: "The credit goes more to behaviorism than to psychoanalysis of course. Academic psychology was already becoming more functional, and the birth of behavior theory provided a new potential partner for psychoanalysis" (1992: 208). Despite the efforts of various persons, especially Clark Hull at Yale, to verify psychoanalytic observations and theoretical constructs by so-called objective methods, attempts to operationalize psychoanalytic concepts proved largely unsuccessful (not least because psychoanalysis was "mentalistic," whereas behaviorism was "physicalistic"). Thus, Sears laments the missed opportunity: "What if Freud had been more open to the heterodoxy of Jung, Adler, Rank, and Ferenczi? What if Watson and Lashley had been responsive to the substantive variables of psychoanalysis instead of being hostile to its necessary concern with consciousness and the unconscious?" (1992: 219).

From its inception in the mid-nineteenth century, the discipline of psychology found itself suspended between a biotropic pole that emphasized the experimental study of the "brain" (i.e., the locus of behavior and response) and a sociotropic pole that emphasized the clinical analysis of the "mind" (i.e., the locus of volition, will, and desire). It seems apparent, however, that there existed one underlying accord between the two positions—the desire to make psychology an autonomous and systematic discipline. Furthermore, it is clear that there existed a widespread desire to bridge the gap between the two tendencies. Why did the two poles of psychological research remain incommensurable? In effect, psychology became the playing field for the conflicting impulses of the two cultures: on the one hand, structuralists, functionalists, and eventually behaviorists borrowed the methods of the natural sciences in an effort to explain the operations of the brain; on the other hand, psychoanalysts and humanist psychologists were forced to devise nonexperimental methods to elucidate the workings of the mind. Consequently, the antinomy at the heart of psychological research crystalized in the tension between two knowledge movements: behaviorism and psychoanalysis. By the second decade of the twentieth century, behaviorism had emerged as the representative of the biotropic or experimental tendency, while psychoanalysis had established itself as the emissary of the sociotropic or clinical tendency.

Notes

1. Seeking to overcome the antagonism between the biotropic and sociotropic approaches, Miller emphasizes the need to restore "immediate experience" as the

subject matter of psychology: "Some parts of our explanation would depend on the neurophysiology of receptors, brain, and effectors; some parts would depend upon the social conventions that civilized people must internalize. But an explanation of consciousness could not be cannibalized by either biology or sociology." Thus, Miller appeals to "a faith that somehow, someday, someone will create a science of 'immediate experience'" (1992: 43–44, 42).

2. In Pribram's opinion, "the advent of behaviorism should have immediately altered our views on the privacy of self-experience, the privacy of perception, thought, and feeling. Though not directly accessible to others, self-experiences can be verbally reported, consensually validated and in this fashion made 'objective.'" However, behaviorism excluded all subjective terminology, thereby "leaving one with a clean behavioral science [that] overlaps with but does not cover the range of a psychological science" (1992: 710–11).

3. As Amedeo Giorgi has noted, "most of the difficulties cited by psychologists concerning the disciplinary status of psychology can be grouped under three headings: (1) the lack of unity in psychology; (2) the irreconcilable split between the scientific and professional aspirations of the field; (3) the apparent discrepancy between psychology's commitment to be scientific and its ability to be faithful to either the givens of the human person or the characteristics of concrete phenomena" (1992: 49).

4. It is fitting, therefore, that the celebration of the centenary of Wilhelm Wundt's laboratory at Leipzig, which the American Psychological Association has posited as the founding gesture of systematic psychology, provided the occasion for a dispute over the definition of the discipline. To mark the event, Sigmund Koch, a prominent figure in the American Psychological Association, offered the following reflections: "What Wundt effectuated by his 'founding' was a semantic change. . . . Henceforward the core meaning of 'psychology' would be dominated by the adjectives scientific and experimental" (Koch 1992: 8–9).

Interestingly enough, Wundt was not cognizant of the significance of the founding gesture. As Koch has mentioned, "Wundt had been assigned a small room at Leipzig University for the storage of demonstration apparatus as early as 1876. During the course of his psychological seminar in the winter of 1879, two of his students—Max Friedrich and Stanley Hall [who would later make a decisive contribution to the development of experimental psychology in the United States]—began to use that space for research: Friedrich as investigator, and Hall as occasional observer" (Koch 1992: 8–9). It was not until 1882 that the space became a full-fledged laboratory.

5. As Joseph Ben-David and Randall Collins (1966) have suggested, the conflict between the disciplines of philosophy and physiology established the institutional precondition for the emergence of psychological research. It is worth noting, therefore, that the originators of systematic psychology—Wundt, James, and Freud—were philosophy-physiology role hybrids. In late nineteenth-century Germany, three elements contributed to role-hybridization: "(a) an academic rather than an amateur role for both philosophers and physiologists; (b) a better competitive situation in philosophy than in physiology encouraging the mobility of men and methods into philosophy; (c) an academic standing of philosophy below that of physiology, requiring the physiologist to maintain his scientific standing by applying empirical methods to the materials of philosophy" (Ben-David & Collins 1966: 465). Doubtless, the situation in the United States was somewhat different. The new psychology encountered considerably less resistance from the philosophical establishment; and experimental psychology received ample support from the new universities (Clark, Johns Hopkins, and Chicago). Nonetheless, it is clear that James remained ambivalent about the prospects for experimental psychology. On the one hand, he recognized the need for

physiological research, and on the other, he lamented the growing opposition to "philosophical speculation."

6. Following E. B. Titchener, Boring believed that every experimental psychologist should be well versed in the history of the discipline (or, more precisely, that historical knowledge was a prerequisite for sensible experimentation): "Every psychologist who was trained for the doctorate in E. B. Titchener's Psychological Laboratory at Cornell in the first two decades of this century knew—or believed—that the historical approach to the understanding of scientific fact is what differentiated the scholar in science from the mere experimenter" (Boring 1961: 3). Similarly, in advocating "contentual and methodological pluralism in psychology and in science," Sigmund Koch harbored a preference "for the restoration of man as an object of serious, direct, and unembarrassed inquiry within certain segments of psychological science" (Koch 1959: 1:2).

7. Wundt's most renowned student, E. B. Titchener, defined experimental psychology as "the study of the basic units of human conscious experience, analyzed through the use of experimental and introspective procedures." Interestingly enough, "topics outside Titchener's definition of 'experimental' were, for example, mental testing, child study, abnormal psychology, and animal psychology" (Goodwin 1985: 384). It is not surprising, therefore, that Titchener eventually found himself at odds not only with psychoanalysis (with its emphasis upon unconscious processes and abnormal psychological states), but also with behaviorism (with its debt to animal psychology).

8. The structuralist and functionalist movements, which received their names in Titchener's 1899 essay "Structural and Functional Psychology," were largely contemporaneous. Functionalism constituted the breeding ground for behaviorism (especially at the University of Chicago). For its part, behaviorism, though operative as a tendency in animal psychology and comparative psychology for many years, received its first explicit formulation in Watson's programmatic essays of the 1910s. By the time Watson emerged at the forefront of psychological research, the debate between structuralism and functionalism had long since lost its significance, due both to the collapse of structuralism owing to the refutation of its fundamental postulates, and to the incorporation of functionalism by behaviorism.

9. As Koch noted, Tolman made a significant contribution to dynamic psychology in the 1930s in the form of the intervening variable paradigm. Introduced "as a modest device for illustrating how analogs to the subjectivists' 'mental processes' might be objectively defined, the concept of the intervening variable was soon thereafter elaborated by Tolman and others [most notably Clark Hull], into a paradigm purporting to exhibit the arrangement of variables which must obtain in any psychological theory seeking reasonable explanatory generality and economy" (1973: 698).

10. In *Elemente der Psychophysik*, Gustav Fechner attempted to establish an "exact science of the functional relations or relations of dependency between body and mind." Though renowned as a precursor of scientific psychology, Fechner was ambivalent about the application of scientific techniques in psychological research. Thus, as Boring notes in the editor's introduction to the translation of his influential book, "[Fechner] might have been content to let experimental psychology as an independent science remain in the womb of time, could he but have established his spiritualistic 'day view' of reality as a substitute for the current materialistic 'night view.' The world, however, chose for him; it seized upon the psychophysical experiments, which Fechner meant merely as contributory to his philosophy, and made them into an experimental psychology" ([1860] 1966: xv, xi). Thus, Fechner contributed unwittingly to the burgeoning "anti-philosophy" movement. Leading the experimentalist assault on the

faculty of philosophy, Hermann von Helmholtz (1821–1894) and his associates argued that "no forces other than the common physical-chemical ones are active within the organism" (Fancher 1979: 95). As an assistant to Helmholtz, Wilhelm Wundt conducted experiments on the nervous system and the sense organs. His early participation in the antiphilosophical movement notwithstanding, Wundt became a professor of philosophy at Leipzig in 1879.

11. Koch attempted to redress the deficiencies of Snow's analysis: "[Snow] leaves psychology and social sciences out of the picture and thereby, I think, effects a serious distortion. For, one of the unique features of psychology is precisely that this is an area in which the two cultures must be in contact" (1961: 636ff).

12. The widespread desire for a truly scientific psychology fueled the growth of the discipline (especially in the period 1880–1920). In Germany, the number of serial publications in scientific psychology (as distinguished from psychoanalysis and psychiatry) increased from one in 1880 to twenty-six in 1920; the number of institutes and laboratories devoted to scientific psychology increased from one in 1880 to nineteen in 1920 (Kusch 1995: 123–25). Moreover, by the second decade of the twentieth century (i.e., the behaviorist revolution), the United States had achieved "a massive hegemony, relative to the rest of the world, in the size of its psychological workforce; the number of its psychological laboratories and scholarly periodicals; the number of university and other contexts in which psychology was pursued; and in the amount of public interest, both formal-educational and lay, which the field commanded" (Koch 1992: 22).

13. The University of Chicago, which served as the seat of functionalist (and eventually behaviorist) psychology from the turn of the century until the First World War, was the site of many attempts to make psychology the infrastructure of the social sciences. The influence of the Chicago school, which embraced pragmatism in philosophy and functionalism in psychology, was felt in economics, political science, and especially sociology. Thus, as John Mills notes, the Chicago sociologists demonstrated "the formative and continuing role of Progressivism and the convergence of that heritage with a behaviorist positivism." For its part, the neobehaviorist strain of positivism, which derived from the grafting of the Vienna Circle's project for a unified science onto the insights of Watson, "provided a minimal, agreed set of standards for the conduct of research and teaching," thereby projecting "the reassuring image of groups of scholars pursuing objective, disinterested research and then offering their findings to society" (J. Mills 1998: 28, 26).

14. Mackenzie isolates two crucial features of behaviorism: "The first was the repudiation of unobservable entities and processes—particularly the mind and consciousness, but by extension others as well. . . . [T]he second was the adherence to explicit decision procedures as the basis for evaluating scientific statements, concepts and theories" (1977: 54).

15. Freud received his "scientific education at the Institute of Ernst Brucke, the Professor of Physiology at the University of Vienna," where he was exposed "to the ideas of Helmholtz and, less directly, Fechner—the twin pillars of the arch by which students of physiology entered Wundt's laboratory" (Rosenzweig 1992a: 136–37). There seems little doubt that Freud's background at Brucke's laboratory contributed to his initial belief in the scientificity of psychoanalysis.

16. As the first American to study under Wundt at Leipzig, the first person to receive a doctorate in psychology from Harvard University (under the tutelage of William James), the first president of Clark University, and the founder of the American Psychological Association (1892), G. Stanley Hall exerted considerable influence on the development of American psychology.

17. With a view to reconciling the various tendencies in psychological research, G. Stanley Hall managed to "[induce] Sigmund Freud of Vienna, W. Stern of Breslau, C. G. Jung of Zurich, E. B. Titchener of Cornell, F. Boas of Columbia, Adolph Meyer of the Johns Hopkins University (both the latter formerly at Clark), H. S. Jennings of Hopkins, H. Ferenczi of Prague, Ernest Jones of Toronto, and William James, to be present and speak" (G. Hall 1923: 332–33).

18. Hall was not alone in believing that the findings of psychoanalysis could be confirmed through experimentation. In fact, as Rosenzweig has noted, "the decade or two following the Clark conference witnessed sporadic efforts to subject aspects of psychoanalytic theory to the test of experimental verification" (Rosenzweig 1992b: 158).

19. Putnam, who was elected president of the American Psychoanalytic Association at its inaugural conference in 1911, defended psychoanalysis in a series of debates with Morton Prince and the Boston School of Psychopathology (Rosenzweig 1992a: 158).

✳

6

Orientalism and Area Studies:
The Case of Sinology

Ho-fung Hung

As the primary source of information on non-Western societies in the
nineteenth and early twentieth centuries, Orientalist scholarship was
crucial to the formation of the social scientific enterprises that attempted
to create universal theories out of the examination of Western societies,
always with an implicit comparison to non-Western ones. Recent critics
of Orientalism (and of area studies), however, have argued that it has
been complicit with colonial powers insofar as they ossified the difference between Western and non-Western civilizations and suggested the
natural superiority of the former. Orientalism is portrayed as a discourse
operating on two fundamental premises unchanged in the last two hundred years. The first is that each non-Western civilization is a unified
whole with some essential characteristics that distinguish it from all
other civilizations. The second is that non-Western people are incapable
of studying their own cultures scientifically. Hence the study of the Orient is a responsibility, as well as the privilege, of Western scholars (Abdel-Malek 1963; Said 1978; B. Turner 1978).

Despite the pertinence of the critique, it takes insufficient account of
the twists and turns of Western representations of the East, and the
complicated relations between Oriental studies and other academic

disciplines. Orientalist discourses have varied over time and space as a function of the dynamics of both the modern world-system and its structures of knowledge. The trajectory of Orientalism has been shaped by the continuing tensions between the two cultures.

In the analysis of the formation and transformation of Orientalism between circa 1800 and 1960, we shall utilize Western Sinology as an exemplary case. China differs from India and the Middle East in that it is located farther from Europe and it was never colonized outright. Hence Sinology is a branch of Orientalism that has perhaps been somewhat more sensitive to the changes in the Western conception of the East.[1]

European enthusiasm for China may be traced at least to the sixteenth century, when the Jesuits began their conversion campaign in China. Under what was called the "accommodation policy," the Jesuit strategy was to ally with the state elite in China and combine the teachings of the Roman Catholic Church with indigenous religious ideas. Recognizing that Confucianism was the "state religion" of China, the Jesuits studied Confucian texts intensively and declared that the belief in a monotheistic God was inherent in the teachings of Confucius. The Jesuit interpretation came to be canonized as the authentic version of Chinese culture, other religious patterns in China were disregarded, and Confucian doctrines became known to the European public through Jesuit translations (Dirlik 1996; Rule 1986; Jensen 1997).

European zealotry for China reached its apogee during the seventeenth and early eighteenth centuries in the form of secular Sinophilism among Enlightenment thinkers. Sinophilism was fed by a European identity crisis brought on by the demise of the hegemony of the Roman Catholic Church. Many European intellectuals saw China as a model of great stability and prosperity (in contrast with the chaos in Europe), one achieved without a central role for priests. This image of China thus became a powerful weapon of early Enlightenment thinkers in their challenges to the church. Their call for learning from China was such that it was even said that Confucius became the "patron-saint of the Enlightenment" (Reichwein 1925: 77).

China was represented as a superior civilization in all aspects. Gottfried Wilhelm Leibniz claimed that the most fundamental principles of natural science and mathematics were to be found in the ancient Chinese text *I-Ching* (Mungello 1985: 312–28). John Webb, a philologist who spent most of his life in search of the *lingua adamica*, entitled his 1669 book *An Historical Essay Endeavouring a Probability that the Language of the Empire of China is the Primitive Language*. He argued that since Chinese was the primitive language, it should be used as a universal language for the whole world (Mungello 1985: 183–97; see also Lach 1977: 501–8). Quesnay and his physiocratic followers argued that the Chinese empire represented the most perfect economy, natural and laissez-faire, based on the encouragement of agriculture, simplicity of taxation, and government nonintervention in trade. Quesnay was posthumously called "the Confucius of Europe" by his disciples (Reichwein 1925: 104–7; see also Rowbotham 1942: 286). Voltaire, the most ardent Sinophile of all, wrote approvingly: "[I]f there has ever been a state in which the life, honor and welfare of men has been protected by laws, it is the empire of China" (quoted in W. Davis 1983: 544; see also Leites 1978: 143–51; Guy 1963: 253, 256).

After the 1750s, however, Europe began to consider itself a "progressive continent" (M. Bernal 1987: 204; see also Adas 1989: ch. 2) in the context of renewed geographic and economic expansion of the European world-system (Wallerstein 1989a). European arrogance returned. Racist views of China replaced respect and admiration and became the dominant attitude toward the East in general while Sinophilism came under fierce attack from the radical philosophers. Giambattista Vico in his *New Science* claimed that Confucian philosophy was "rude and clumsy" and almost entirely devoted to "a vulgar morality" (quoted in Zhang 1988: 116). Immanuel Kant wrote, "philosophy is not to be found in the whole Orient. . . . [The] concept of virtue and morality never entered the head of the Chinese" (quoted in Ching 1978: 169). Confucius's writings were denounced by G. W. F. Hegel as "highly tasteless prescriptions for cult and manners" (quoted in Y. Kim 1978: 174). Chinese government was regarded by Rousseau as contradictory to the idea of human progress and as a mere tyranny, ruling by sticks and terror (Guy 1956). According to the pioneers of scientific linguistics, the Chinese language was a "dead language" as opposed to the "living" European languages (Rowbotham 1942: 288).[2]

Europe's enthusiasm for the East disappeared altogether at the end of the eighteenth century when the tornado of the French Revolution swept the continent. When attention to the Orient revived in the early nineteenth century during the "Oriental Renaissance" (Schwab 1984), it led directly to the emergence of Orientalism as a discipline. The renewed attention to the East was, first of all, a consequence of accelerated European colonial expansion and the growing practical need for training diplomats, colonial officials, and translators. It was, second, a result of the romantic movement. The romantics, with a thirst for the mystical, were devoted to the search for a childlike innocence and purity and for a reunification of religion, philosophy, and art that had been sundered in the modern Western world, but whose unity was thought to be preserved in the supposedly stagnant oriental civilizations (J. Clarke 1997: 55–56; M. Bernal 1987: 19–20).[3]

Assuming that people in the Orient were incapable of studying their civilizations systematically and that these civilizations had witnessed some kind of golden age in the past but had been declining thereafter, the romantic Orientalists presumed their own responsibility to study them before they were lost (Said 1978: 113–23; M. Bernal 1987: 235). The resistance of romanticism to the Enlightenment conferred on early Orientalist scholarship an identity in opposition to modernity and the scientific disciplines. Given the enthusiasm for spirituality within the romantic movement, romantic Orientalism reduced all non-Western civilizations to different world religions. This was in sharp contrast with Europeans' broad interests in the political economy, technology, and the philosophy of the East in previous centuries. With now the assumption that oriental cultures never changed, analysis of ancient texts and linguistic structures (also supposedly unchanging) became the main tool in the field.

Although Orientalism tended to be idiographic, Orientalists claimed to use "scientific methods"—understood to mean systemic analysis—as the means to unveil and classify oriental civilizations. This was achieved by the canonization of the oriental "classic texts" and the construction of a taxonomy of oriental cultures. Over the nineteenth century, this process was crystalized in the monumental fifty-volume series, *The Sacred Books of the East,* compiled by Friedrich Max Müller, a German authority in comparative religion and philology who taught at Oxford. The compilation was conceived so as to embrace all of the essential classical texts of Oriental civilizations.

The presuppositions of the compilation are to be found in Müller's theory of comparative religion, which dominated the early Orientalist enterprise. The basic idea was that human history was a history of degeneration and the struggle against it. All religions developed from a common root, which was primitive, natural, rational, and monotheistic. Müller emphasized the commonality of all world religions; he stated that "if we all but listen attentively, we can hear in all religions a groaning of the spirit . . . a longing after the Infinite, a love of God" (quoted in Trompf 1978: 70). It followed that to discover God's purpose one had to study not one but all the religions of the world, especially the religions in their most ancient forms. Müller, himself a major figure in the romantic movement, believed that there was something pure and superior in ancient religious systems that had been lost during the course of cultural development. Studying the supposedly stagnant cultures of the East was the key to rediscovering this glamorous purity. In *The Sacred Books of the East,* Müller himself translated the ancient Indian text the *Rig-Veda.* In it he claimed to perceive a religion in a state of primitive purity, one that possessed conceptions of the Deity "purer and better" than the "savage communities of today" (quoted in Trompf 1978: 71).

In line with Müller's search for a common root of religion was a prominent philological current, which proposed that all human languages shared a single origin. Gustaaf Schlegel, Joseph Edkins, and August Schleicher were representative figures of the movement. They theorized two successive phases of development of human language. The first phase was a progressive evolution from monosyllables to fully developed inflection. The second phase was a decline from the climax of the first phase. Some of them suggested that English and other modern European languages were in the declining phase, whereas Chinese might be in a more primitive stage of development. Hence, Chinese language might not yet have entered the phase of decline and could therefore be the key to deciphering the common human language in its highest form (Pulleyblank 1995: 343–44).

These two theories underpinned the general discourse of romantic Orientalism.

Institutionally, chairs of Oriental studies were established in universities all over Europe. Professional organizations (such as the Royal Asiatic Society in Great Britain) were established one after the other. In 1873, representatives of the major Orientalist institutions met in Paris and founded the International Congress of Orientalists, which met every two or three years thereafter. Each congress was divided into sections according to geographical zones. The purpose was to let the specialists of each zone (the Middle East, India, China, etc.) meet and become acquainted with one another's work, as well as the latest trends within the profession as a whole. It marked the full institutionalization of Orientalism as a unified field of knowledge production.

The Formation and Transformation of European Sinology: 1800–1945

The first chair of Chinese language and literature was established in 1814 at the Collège de France. Sinology as a discipline treated Taoism and Confucianism as the two axes of Chinese civilization. *The Texts of Confucianism* and *The Texts of Taoism* (both translated by James Legge) constituted the volumes on China in *The Sacred Books of the East*. Whereas Confucianism had already been raised to canonical status in the sixteenth century by the Jesuits, Taoism was "discovered" or "invented" by the nineteenth-century Sinologists.

In the nineteenth century, European enthusiasm abruptly turned from Confucianism to Taoism. In fact, the term "Taoism" was a European invention. Taoism is "a rare example of a subject related to Chinese culture which has, in modern times, received perhaps more attention abroad than in China itself" (Schipper 1995: 467). Most founders of nineteenth-century Sinology are remembered more or less in terms of their contribution to introducing Taoism to the West (Schipper 1995: 467).

In temples and monasteries of local religious sects, Lao-tzu, who was thought to be a contemporary of Confucius, was worshiped alongside a number of spirits and gods. *Tao Te Ching,* the book supposedly written by him, was studied by the priests who propagated beliefs in supernatural forces. Those sects were diversified in practice and style, and they drew their adepts from the lower ranks of Chinese society. But they were sanctioned by the Confucian elite. Though they were mentioned occasionally in the writings of the Jesuits and other European travelers long before the nineteenth century, they had never been invoked with as much zeal by Sinophiles as Confucianism. Nor had anybody before tried to link the ancient text of *Tao Te Ching* with the living practices of the priests and present a unified and legitimate Taoist tradition in China.

Jean-Pierre Abel-Remusat, the French Sinologist who was the first professor of Chinese language and literature at the Collège de France, was among the first to call for serious attention to *Tao Te Ching* as a Chinese classic text. He considered that it incarnated a major philosophical/religious doctrine and urged its translation into European languages.[4] James Legge accomplished the apotheosis of Taoism when he published *The Texts of Taoism* as volumes 39 and 40 of Müller's *Sacred Books of the East*.[5]

The themes pursued by romantic Sinology can most clearly be found in the works of Legge and J. J. M. de Groot, both exemplars of the study of Chinese religions and, respectively, the leading figures of the tradition of textual criticism and of ethnography in Sinology.

Legge was a missionary from England who worked in Hong Kong, Macao, and Canton from 1843 to 1873. In 1861, he published his translation of the Confucian classics as the series Chinese Classics. It was reputed to be the most authoritative translation of the works of Confucius and his disciples (Ride 1960: 20–21). Later, as we have noted, Legge turned his attention to Taoism as well. In 1876, a chair of Chinese language was established in Oxford University and Legge was appointed its first occupant.

Legge's understanding of Chinese religions took shape in his early career as a missionary (H. Legge 1905: 68). His interpretation of Chinese religions was systematically expressed in his lecture given to a group of English missionaries in 1880, "The Religions of China: Confucianism and Taoism." He asserted that both Confucianism and Taoism, in their most ancient and purest form, were monotheistic. Buddhism was only mentioned briefly and treated as an external import that caused the corruption of the other two.

Confucianism and Taoism were compared with Christianity in the last section of the lecture. Legge argued that the three religions agreed with one another on three fundamental points. First, all of them believed in a single God. Second, the idea of the possibility and of the fact of revelation was common to them. And last, all three believed in the supernatural (J. Legge 1976 [1880]: 245–48). In general, Legge was sympathetic to Chinese culture and treated it as an equal with Christianity:

> [The] divine stamp on Christianity must not be supposed to imprint the brand of falsehood on other religions. They are still to be tested

according to what they are in themselves. . . . It must be borne in mind also that when we have concluded that Christianity is the revealed religion, this does not relieve us from the task of searching the scriptures diligently, and finding out their meaning by all legitimate methods of criticism and interpretation. The books of the Old and New Testaments have come down to us just as the Greek and Roman and Chinese classics have done, exposed in the same way to corruption and alteration, to additions and mutilations. The text of them all has to be settled by the same canons of criticism. . . . During my long residence among the Chinese, I learned to think more highly of them than many of our countrymen do; more highly as to their actual capacity, and more highly as to their intellectual capacity. (J. Legge [1880] 1976: 286–88, 308)

Legge and Müller had known each other from 1875, and it was under Müller's recommendation and encouragement that Legge decided to teach at Oxford. In fact, Legge's views toward China were very much influenced by the works of Müller, as reflected in the introductory part of *The Religions of China*, in which Müller's scheme of comparative philology and religion was strictly followed. Legge's "discovery" of God in his reading of ancient Chinese texts echoed Müller's theory of primitive religion.

Legge's positive attitudes toward China were shared by de Groot. De Groot was trained at Leiden University and was a student of Gustaaf Schlegel, one of the founding fathers of European Sinology and, as of 1876, the first professor of Chinese language and literature at Leiden. After finishing his studies, de Groot joined the Dutch colonial services and worked in Amoy (Xiamen) in 1877–1878 and 1886–1890. There he conducted ethnographic observation of popular religious practices and compiled his findings into *Les fêtes annuellement célébrées à Emoui (Amoy): Étude concernant la religion populaire des Chinois* (1886). He returned to the Netherlands in 1890 and turned to an academic career. In 1891, he was appointed as the professor of ethnology of the Dutch East Indies at Leiden University. Then he published the six volumes of *The Religious System of China: Its Ancient Forms, Evolution, History and Present Aspect, Manners, Customs and Social Institutions Connected Therewith* (1892–1910), which was considered a "monumental" study of Chinese popular religion and Taoism (Freedman 1979: 356; Schipper 1995: 473). After the death of Schlegel, he took up the chair of Chinese language in 1904. In 1912, he accepted an appointment from the University of Berlin and became the professor of Chinese there. In the same year, he published *Religion in China: Universism—A Key to the Study of Taoism and Confucianism*, which was an ambitious attempt to give a comprehensive account of Chinese official and popular religions, so as to provide "a key to the study of Taoism and Confucianism" (de Groot 1912: vi).

De Groot is regarded as one of the most important "Sinological-sociological" contributors to the study of Chinese religion (Freedman

1979: 355) and "the first to use the sociological methodology which has exerted such an important influence on the study of Chinese religion in the 20th century" (Schipper 1995: 472). His attitude toward China was particularly sympathetic (an outlook supposedly inherited from Schlegel); of his earlier book, Maurice Freedman remarks:

> The picture painted of China in Les fêtes is complimentary (more so in the original Dutch than in the French translation). China is an alternative civilization, having roots in common with Europe. It is to be compared with Europe, in some respects very favorably. The book expresses anti-Christian, especially anti-Catholic, sentiments, and emphasizes the religious tolerance prevailing in China. . . . Spencerian evolution of a sort is present in his earliest work, but are [sic] countered by an eighteenth-century respect for China. (Freedman 1979: 357)

Both Legge and de Groot wrote in a romantic tone, with a kind of universal humanism. Nonetheless, at the end of the nineteenth century, they underwent a simultaneous reversal of their attitudes toward China.

By that time, Legge had totally shifted his interest from Confucianism to Taoism and had abandoned his original ideas that Chinese religions were monotheistic in nature. In the introduction to *The Texts of Taoism* (1891), Legge tried hard to construct a superstitious image of Taoism. The Taoist doctrines were found to be grounded on an "agnosticism to God" and the magical "belief that by certain management and discipline of the breath life might be prolonged indefinitely" ([1891] 1957: 87). The Taoist texts were "unbelievable, often grotesque and absurd" ([1891] 1957: 85). Moreover, "[a] visitor to one of the larger of these [Taoist] temples may not only see the pictures of the purgatorial courts and other forms of the modern superstitions, but he will find also astrologers, diviners, geomancers, physiognomists . . ." ([1891] 1957: 90). Above all, "Taoism was wrong in its opposition to the increase of knowledge," and "[w]e can laugh at it," as "[m]an exists under a law of progress" ([1891] 1957: 75). It is surprising that Legge, originally a romantic, would finally come to believe in the "law of progress."

De Groot underwent a similar radical change over the same period. Hostility toward China was expressed everywhere in his influential later works, the six volumes of *The Religious System of China* (1892–1910) and *Religion in China: Universism* (1912). Though de Groot did not have a social science background and his works came before the publication of Émile Durkheim's *Elementary Forms of the Religious Life*, his general methodology placed him in the same movement of sociological and anthropological study of non-Western religions under an evolutionary paradigm (Schipper 1995: 472; Freedman 1979: 355–61). In this paradigm, "primitive religions" in the non-Western world lost all advantages over Western civilization and were seen as simple animism and fetishism. This was fundamentally different from Müller's model of comparative religions that placed the so-called primitive religions on a higher plane than the modern ones.[6]

The Religious System of China was a study of Taoist rituals and customs. It was perceived by the author as the first attempt to introduce "science" into the study of Chinese religions: "the author has followed the beaten track of Science for the study of Religions and Sociology in general" (de Groot [1892] 1969: ix, xi–xii). But, in the eyes of de Groot, the Chinese were one of the "semi-civilized peoples" who were traditionalistic and resistant to change. In China, one could find that "[m]any [ancient] rites and practices still flourish among the Chinese, which one would scarcely expect to find anywhere except amongst savages in a low stage of culture" (de Groot [1892] 1969: x, xi).

De Groot's hostility toward China was even more explicit in *Religion in China: Universism*. In it, he ambitiously attempted to outline the unifying principle of Chinese religions. At the outset, the author asserted that three religions of China, Confucianism, Taoism and Buddhism, were all the same: "the religion of the Universe, its parts and phenomena. Its Universism . . . is the one religion of China" (de Groot 1912: 3). Of the three, Buddhism was left aside, as it was "merely the engrafted branch" (1912: 3). Between Taoism and Confucianism, the former was more fundamental, as "Universism was Taoism; the two terms are synonymous" (1912: 3). The reduction of different traditions into one allowed him to describe the essence of Chinese culture by just focusing on the Taoist texts and the practices of folk religions supposedly rooted in those texts.

De Groot asserted that the Chinese worshiped the animated Universe as a mystical force and also worshiped all spirits, gods, and ghosts, which were parts of this Universe. Taoists worshiped spirits in rocks, hills, and rivers. Confucians worshiped dead men and their ghosts. Thus, China was "the principal idolatrous and fetish-worshipping country in the world" (1912: 188). All the ancient classics of China were texts teaching rituals. De Groot echoed Legge and stated that Taoism "suppress[es] knowledge and wisdom" (1912: 65–66) and drew the same conclusions as regards the "Universalism which we call Confucianism" (1912: 189). De Groot's views here were essentially different from those expressed in his earlier work. A radical break of viewpoint must have occurred somewhere between 1886 and 1892.[7]

The transformations of Legge and de Groot's views on China were strikingly coincidental. They occurred in 1883–1891 and 1886–1892, respectively. It has been argued that the reason for the change in Legge's disposition was the "public criticism engendered by the new science of religion," which called for "a 'more scholarlike spirit' that did not bow to either the unfounded fears or foolish enthusiasms of the non-academic public over 'Oriental Wisdom'" at the end of the nineteenth century (Girardot 1992: 190). Freedman believes that de Groot's "manner of presenting China after his appointment to Leiden reflected the anthropology of the day" and says, "One might well wish that he had kept away from Science and been faithful to the humanism of his early work" (1979: 357–58).

But this change in viewpoints is not best explained biographically. Rather, it represented a general turn in the epistemological foundation of the Orientalist enterprise:

> [There] is the passage from a kind of late, idealistically inclined, nineteenth-century "integral humanism" or "hermeneutics of trust" (exemplified by the innate historical and literary piety of Legge's Sinology and Müller's science of religion) to a more highly specialized, rationalistic, and secularly academic "hermeneutics of suspicion" concerning the integrity of ancient history, texts, and authors. . . . In many ways this issue comes down to the difference between those holding onto the "enormous antiquity," "purity," "sacredness," or "authenticity" of Oriental texts versus a "higher" critical attitude that demolishes "ancient authority," "sacred books," and "religion" into so many disparate historical and philological fragments. (Girardot 1992: 190)

Indeed, the transformation of Oriental scholarship was part of the transformation of the general intellectual atmosphere in the period. In the four decades between the Berlin Congress and the First World War (1878–1914) Europe witnessed the greatest colonial expansion in history. In the case of China, the 1890s marked an upsurge of European penetration of the empire. Between the Opium War of 1842 and the defeat of China in the Sino-Japanese War in 1895, the Europeans had established only a few footholds at the coastal treaty ports. Afterwards, the imperialist powers were in a rush to divide China among themselves and encircle their "spheres of influence." In this context, romanticism faded out and social Darwinism came to the fore. The romantic respect of the Orient gave way to scientific racism, and the idea of progress was driven to the extreme (see Leaf 1979: 106–39; Driver 1992: 25–27; Hobsbawn 1989: 253).

The impact of social Darwinism on Oriental studies may be seen in Müller's inaugural speech as president of the Ninth International Congress of Orientalists in 1892. By then, Müller no longer was the universalist humanist. The name of Darwin was invoked repeatedly, and the essential difference between the "Aryans" and the "dark men" was seen as the fundamental premise of Oriental studies.[8] The scientific-racist turn of Orientalism marked the end of the romantic resistance to modernity in the discipline, which had been pushed to the scientific pole of the two cultures.

Legge's and de Groot's later works were the bases of Max Weber's discussion of Chinese religions in *The Religion of China: Confucianism and Taoism* ([1922] 1951). Legge's translations and interpretations of the Confucian and Taoist texts, together with de Groot's analysis of Chinese folk religions, constituted the most significant sources for Weber's knowledge about Chinese culture.[9] Basically, Weber's description of Confucianism and Taoism (as well as of Buddhism seen as an imported and Taoicized religion) was a replication of Legge's and de Groot's perspectives and vocabularies. According to Weber, the worldview of the Chinese, Confucianist literati and Taoist laymen alike, was "magical" and based on "a crude, abstruse, universist conception of the unity of the world," in addition to an "animist compulsion of spirits." In China, even "the high Chinese officials . . . did not hesitate to be edified by the stupidest miracle." These attitudes were "strongly counteractive to capitalist development," and "a rational economy and technology of modern occidental character was simply out of the question" ([1922] 1951: 227, 229, 249, 227).

Although Weber's writings on China and other non-Western cultures were not widely read, they were important building blocks of his comparative religion studies and his thesis on Protestant ethics. Weber's theory on culture and capitalist development prepared the ground for Parsonian social science and the modernization school that flourished in the postwar United States. It provided the impetus for Talcott Parsons's idea, as well as that of modernization theorists, that all non-Western traditional cultures were impediments to modernization. The scientific-racist turn of Legge and de Groot under the strong push of science, and their influence on the development of social science theories in return, illustrates the interactive relation between Oriental studies and the social science disciplines dealing with the Western world.

Sinology after 1945

After the traumas of fascism and two world wars, social Darwinism ceased to be in the mainstream of the European academic world. In this atmosphere, European Sinology regained its identity as a humanistic study and returned to its pre-1890 romantic tone. Admiration of Chinese antiquity once again became the ethos of the Sinologists. The romantic underpinnings of the discipline—that studying the Orient was essential for complementing the study of Western modernity, and that such study was a responsibility of Western scholars for the world's sake—reappeared. Descriptions of the field like the following were not difficult to find:

> Why should we study China? You can use botany as an analogy. Each culture is like a species of plant; it has strong points and weak points different from the others. The same kind of plant will weaken after a while if it's not cross-fertilized to improve the next generation. The botanists of the world keep a gene bank to preserve seeds that may or

may not be needed now, because you never know what problems may come up with varieties that are flourishing now, and they might be needed later. But the first step of cross-fertilization is to preserve a "pure variety" with its own characteristics. So the study of traditional Chinese culture, in the long view, is really good for world culture. . . . Our consensus is: Chinese culture is too serious to be left to the Chinese alone. (Interview with Kristofer Schipper, L. Li 1991: 121)

Meanwhile, however, as the disintegration of the European empires dried up financial resources for Orientalist departments in the universities, European Orientalism was marked by the postwar dissolution of the international congresses of Orientalists. The issue of disbanding the congresses as they "no longer serve[d] a useful purpose" under the new geopolitical situation was first raised among the participants of the twenty-first congress in Paris in 1948 (Société Asiatique de Paris 1949: 37). In 1973, during the celebration of the hundredth anniversary of the congress, a resolution was made to bring the congress to an end. The Project of Reform of the International Congress of Orientalists criticized the "Europocentric aspects of studies of Asia, as is practiced in the Occident in recent generations." The proposal also rejected the "prejudices holding that peoples of Asia were not competent to carry out studies of Asia." As more and more Asians became active in the research on the Asian societies, it was an "absurdity" to label the students of Asian civilization as "Orientalists" (Filliozat 1975: 57).

Corresponding to the decline of European Orientalism was the rise of area studies in the postwar United States. Under the hegemony of Parsonian structural-functionalism, modernization theory as derived from the evolutionary sociology of nineteenth-century Europe, hand in hand with the thirst for practical knowledge of different geographical areas given by the practical needs of the Cold War, shaped the formation of area studies. Institutionally, area studies were embedded in professional associations (such as the Association for Asian Studies founded in 1941, which published the journal *Far Eastern Quarterly*, later the *Journal of Asian Studies*) and university programs supported by foundations and government funding.[10] Norman Brown outlines the trajectory of the founding of area studies:

In the United States it seemed obvious that to win the [Second World War] and the peace afterwards it was necessary to have the cooperation of the Oriental peoples, and the many wartime agencies began to recruit personnel acquainted with the Oriental world. . . . To many persons in the United States during the war years and the years following it was evident that there was a need for the national education system to expand so as to include study of the modern Orient. The great philanthropic foundations took the lead, first the Carnegie Corporation of New York, then the Rockefeller Foundation, and a few years later the Ford Foundation when it was established. The immediate need which the foundations saw was to cultivate oriental studies in the social sciences fields—economics, politics and government, international relations, the behavioral sciences. . . . The United States Congress also in time came to see the importance of such studies and enacted legislation to support them. (1971: 32)

This logic of course applied equally to the Cold War.

Postwar Sinology in the United States was known as China studies. A leading figure was John King Fairbank, who, along with his so-called Harvard school, put the field under the domination of his "impact-response" model (see Fairbank 1953; Fairbank et al. 1965). Informed by modernization theory, the model's basic claim was that modernization in East Asia was impeded by Chinese traditional culture, and the region's modernization could only occur as a response to external stimulus. Impact from outside was the only hope of progress, if any.

The relation between modernization theory and China studies was not one-way. Fairbankian China studies, especially Fairbank's masterpiece, *Trade and Diplomacy on the China Coast* (1953), contributed to the general theory by providing an implicit example of modernizing transformation without colonial violence (see Barlow 1997; Farquhar & Hevia 1993: 486–94). China's treaty ports, especially Shanghai, in the nineteenth and early twentieth centuries were presented as utopian communities representing the promise of a modernized China. The fact that China was never formally colonized let Fairbank depict the modernization process as peaceful, natural, and harmonious. This made China not just one of the ordinary examples of modernization theory. Instead,

the "China" of structural-functionalism is the instantiating trope for the theory generally. Without their version of "China," structuralism and modernization theory would have lacked proof of their claim that cultural transformation could occur "naturally" without revolution, and that would certainly have robbed social science theory of its potential to counter the powerful Marxisms of Lenin, Stalin, and Mao. (Barlow 1997: 393–94)

Postwar China studies was highly nomothetic. This was in contrast with the humanistic identity of postwar European Sinology. However, the impact-response model and modernization theory in general were not totally alien to the European tradition. They could be regarded as a transplantation and variant of late nineteenth-century scientific Oriental studies, as we have already seen how the Weberian sociology of non-Western religions had acted as an intermediate step (see Lele 1993). The temporal discontinuity between romantic Sinology and scientific-racist Sinology in nineteenth- and early twentieth-century Europe was translated into a spatial discontinuity between European Sinology and

U.S. China studies in the second half of the twentieth century, as was indeed observed by the president of the European Sinological Association, who spoke of "the new emerging forces [experts in U.S. China studies] (often trained in political science and economics rather than brought up with the Four Books), accusing traditional Sinologists of being petrified and antiquarian, and Sinologists branding the contemporary China experts as superficial and politicized" (Idema 1991: 9).

The tension between European Sinology and U.S. China studies is in part an expression of the tension between the two cultures.[11] In spite of the links between area studies and the Cold War establishment and Parsonian social science, the institutionalization of area studies led to unintended consequences that sowed the seeds for transforming the Western conception of the East, on the one hand, and posed a challenge to universalist social science on the other. Although area studies were always under the leadership of the social scientists, the fields actually became a locus of interdisciplinary work and for overcoming the "Imperialism of the departments" (Fenton 1947; Palat 1996; Cumings 1998). The institutions of area studies "offered a rare venue where one could see what a historian thought of the work of an economist, or what a literary critic thought of behavioralist sociology" (Cumings 1998: 24). The application of social scientific theory and methods undermined the unity implicitly attributed to the social formations under study and gradually dissolved the Orientalist reduction of non-Western civilizations into oriental religions and texts. In the case of China studies, different branches covering a wide range of areas like kinship, urban system, market structure, demography, foreign trade, and imperial jurisdiction blossomed in the 1950s and 1960s. Religion was no longer the privileged area, and ancient texts no longer the privileged objects of investigation. Simultaneously, area studies brought the specificity of non-Western zones into the vision of the universalist social scientists and created the ground for overcoming Western-centrism in social science. In a speech titled "What the Study of China Can Do for Social Science," William Skinner pointed out that

> [s]ocial scientists . . . have been parochial while claiming universality. They studied Western man and spoke of mankind. . . . If these disciplines are to move away from parochialism toward a universal science of man, we need empirical studies on a large scale in societies outside the Western tradition. . . . All societies are unique, but some are more unique than others. In my view, China is the most unique of all, by which I mean to say that Chinese society has manifested and now manifests so many exceptional features that social scientists ignore it only at the cost of the universality of their science. (1964: 518)

In a similar vein, one of the leading postcolonial critics from India recently traced his critical ideas to the area studies training he received in the United States during the 1960s:

[There are] often noted links between the Cold War, government funding, and university expansion in the organization of area-studies centers after World War II. Nevertheless, area studies has provided the
major counterpoint to the delusions of the view from nowhere that
underwrites much canonical social science . . . [and] one of the few
serious counterweights to the tireless tendency to marginalize huge
parts of the world in the American academy and in American society
more generally. (Appadurai 1996: 16–17)

Institutionally, the rise of area studies changed the "demography" of
faculty in history and other social science departments. The traditional
and Western social science disciplines were "corroded" by area studies
as the number of faculty specializing in non-Western areas in those departments had been increasing since 1945 (Wallerstein 1997a: 219–20). Of
course, the study of non-Western civilizations alone is not a sufficient
condition for undermining the ethnocentric universalism in social science; as we have seen, Weber's Eurocentric sociology was based on his
comparative studies of non-Western societies too. It was not until the
political crisis of 1968 that the radical potentials embedded in area studies were converted into real energy fueling the revalorization of the non-
Western civilizations and contributing to the transformations of the
disciplinary structures.

Notes

1. See Barlow (1997: 400) and Spence (1992, 1998). It is true that the Middle East
was not colonized to the degree that India was in modern times. However, for both
historical and geographical reasons, the West and the Arab world have been in relatively constant contact and conflict over the centuries. It is not surprising that Edward
Said, among other critics, would find that Orientalism has involved constant demon-
ization of the East, generalizing from the case of the Middle East. However, the physical distance between China and Europe has allowed much more space for fantasy
and idealization, permitting the Western conception of China to fluctuate between
admiration and contempt.
2. For a detailed discussion of the development of Western learning on China
before the nineteenth century, see Hung (2003: 254–80).
3. Romantic Orientalism began with the study of India in Germany. It was part of
the project to search for a Germanic cultural identity beyond Enlightenment values
and the French sphere of influence. German Indology was based on the premise that
all European thought originated in Asia, along with a belief in a common Indo-
Germanic Aryan ancestry. It was supposed that the purest roots of the Aryan race
might be found in Indian culture, which was free of "Semitic corruption" (J. Clarke
1997: 55–56, 78). The intellectual current spread rapidly to studies of other Oriental
civilizations in other European countries. Romanticism, as well as romantic Orientalism, was much stronger in Germany, England, and Holland than in France, which
was the epicenter of radical Enlightenment.
4. Nevertheless, the integrity of Taoism as a tradition was constantly challenged
by some Sinologists (e.g., Herbert A. Giles at Cambridge), who argued that Lao-tzu

was a faked figure created in the Han Dynasty (202 BC–AD 8), that the *Tao Te Ching* was a late Han forgery, and that neither of them related to the living practices of the priests. A European consensus on the existence of a unitary Taoist tradition, originating in Lao-tzu's *Tao Te Ching* and perpetuated in popular religious practice, was reached only in the late nineteenth century after a prolonged debate (Girardot 1992).

5. One reason for this sudden creation of, and enthusiasm for, the Taoist tradition at the expense of Confucianism is that the conventional image of Confucianism was out of kilter with the prevalent Orientalist imagery of the dichotomy between "Occident/science/rationalism" and "Orient/spirituality/mysticism." The mystical image of Taoism provided a better fit with the general image of the Orient then prevalent in the West.

6. Durkheim's explicit critique of Müller was the point of departure for his *Elementary Forms of the Religious Life*, in which he argued that all elementary religions were fetishistic and animistic. In the introduction, Durkheim spent quite some efforts to refute Müller's ideas that primitive religions were purer then their contemporary variations (Durkheim [1915] 1976: 71).

7. The views of Marcel Granet, the leading French scholar on Chinese religions before the Second World War, were not that different. His most prominent contribution to the field was to trace all aspects of Chinese religions—Confucianism, Taoism, Buddhism—to the ultimate common root of peasant life and thought, which were supposedly the same from the ancient past to nowadays (see Freedman 1979: 361–62). Though his views on Chinese religions were less negative than that of de Groot, his works echoed, or even extended, de Groot's reductionist view of Chinese cultures.

8. Müller admitted that originally Eastern and Western men were one, but at some point in the prehistoric period, he said, a break had occurred that led to the "great consolidations of the ancient Aryan and Semitic speakers" and a "complete break between East and West." This break "determine[d] the course of the principal nations of ancient history as the mountains determine the course of rivers" (Müller 1893: 9, 15–16).

9. The first Sinologist Weber footnotes in his *The Religion of China* is James Legge. In the chapter "Orthodoxy and Heterodoxy," which deals mainly with Taoism and somewhat with Buddhism, de Groot is quoted extensively. When introducing the sources for Taoism in the chapter's first footnote, Weber writes: "Concerning Taoism, consult de Harlez, and Legge for sources. In general, see the excellent posthumous work of W. Grube, *Religion und Kultur der Chinesen* and especially de Groot's *Universismus: Die Grundlagen der Religion und Ethik, des Staatswesens und der Wissenschaft Chinas*" (M. Weber 1951: 290).

10. Cumings (1998) asserts that the area studies establishment was tightly under the control of what he calls the "state/intelligence/foundation nexus."

11. To be sure, the social scientific approach to the study of China, and the attention to contemporary China, was spreading to Europe. It can be viewed as an attempt of European Sinologists to catch up with America by importing U.S.-style China studies. An example is the establishment of the Documentary and Research Center for Contemporary China at the Sinological Institute at Leiden University in 1968. As recalled by Erik Zurcher, the founder of the center: "The only thing that became quite clear to me [during my stay in the People's Republic of China in 1964] was that my training in classical sinology had not equipped me with the tools to interpret what I saw around me, and that it would be worthwhile, indeed imperative, to widen the scope of Chinese studies at home. . . . Back in Leiden, the response to the idea of drawing contemporary China into the sphere of Oriental Studies was not really enthusiastic. In those years, Asian studies as concentrated in the Faculty of Arts were

going through a period of what may be called 'post-colonial depression.' In the past, most of them had been directly or indirectly related with the colonial administration in the Dutch East Indies, or with the curriculum set up for the training of colonial officials. . . . [A]ll this meant that Oriental Studies were by no means antiquarian or strictly philological. . . . For obvious reasons this had changed in the early 1950s, at least in the Faculty of Arts. Contemporary developments in what came to be called 'the Third World' were largely left to the departments of sociology, anthropology, and political science, and in those areas the University of Amsterdam far outshone Leiden. Then came the 'Ten Years of Turmoil,' starting with the 'Cultural Revolution,' and with the wisdom of hindsight we can say that that cataclysm has given the push needed to add a 'contemporary wing' to the Sinological Institute (and I think that the same happened elsewhere in Europe, where one contemporary China center after another was set up)" (Zurcher 1994: 3–5).

PART II

Contemporary Challenges in and to the Structures of Knowledge

We have seen how over several centuries the concept of the two cultures, the epistemological divide between science and philosophy, has become accepted and institutionalized in the structures of knowledge. In many ways, the period from 1945 through the 1960s represented the culmination of this trend, which started at least four hundred or five hundred years ago and accelerated greatly in the nineteenth century.

Yet, precisely at the moment of its triumph, the concept of the two cultures began to be assailed from many quarters. The attack was not always direct, and the attackers were not always aware that they were attacking, or even interested in, the concept of the two cultures. But the attack was cumulative. We shall, in part II, attempt to lay out the multiple knowledges and social movements that, in one way or another, began to undermine the stability of the epistemological givens of the world-system as they seemed to prevail in the post-1945 period.

*

7

Complexity Studies

Richard E. Lee

As we have seen, the Newtonian model for the study of natural phenomena became solidly established and achieved primacy in the hierarchy of the world of knowledge in the nineteenth century, a claim that
had attained increasing legitimacy from the beginning of the previous
century. Over the course of the nineteenth century, the physical and life
sciences made tremendous gains in the range of phenomena they could
explain and in the sense of certitude that their proclaimed results were
valid. Whether in physics, chemistry, geology, or biology, what now came to
be grouped together simply as the "sciences" passed from being an activity
largely of "amateurs" to a fully professionalized practice carried on by a
cadre of experts denoted by the contemporary neologism "scientist." Ever
more deliberately, the conscious purpose of research was not merely to
observe and explain nature but also to transform it, or at the very least to
harness it—an emphasis, it might be noted, dear to capitalist enterprises.

Toward the end of the century some began to argue that the great
adventure of science was nearing its end, or at least was on the high
road to completing its mission. Electromagnetic theory joined the forces
of light, electricity, and heat. Movement in the Laplacian universe was
fully predictable, even though gravity was not understood. And Dmitry
Ivanovich Mendeleyev had revealed how many elements there were,
and thus how many, or few, were left to be discovered.

However, advances in thermodynamics, by which observed phenomena were placed in direct correlation with one another, put into question mechanical models such as the then current atomic theory and ethereal continua. Theoretical physics began to embrace mathematical models, a trend that "may be dated perhaps to the 1860s when Maxwell abandoned his model of the ether and confined himself to the equations it had given him" (Mason 1962: 502). James Clerk Maxwell and Michael Faraday's concept of the electromagnetic field shook the central-force program that the Newtonian mechanical worldview embodied in physics. Ether theories declined as a result of the experiments of Albert Abraham Michelson and Edward Williams Morley in 1887 and of Sir Oliver Joseph Lodge in 1893. Following the work of George Francis FitzGerald in 1892, the Newtonian model of absolute space (and its ethereal medium) and absolute time (with a single privileged observer) collapsed.

In 1905, Albert Einstein produced a set of papers that were the culmination and consummation of the classical tradition that had begun with Galileo Galilei and Isaac Newton and simultaneously ushered in two great upheavals affecting our view of the physical world. His special theory of relativity showed that the mass of a body was related to its speed (the mass-energy equation $E = mc^2$), and time and space lost their absolute character. This equation showed that a body at high velocity contracted in length in the direction of motion, and time lost its absolute value. Eventually, the general theory showed how a gravitational field is set up as space curves around matter. In the one paper of 1905 that he himself regarded as revolutionary, Einstein contributed to the development of quantum theory, which had originated in Max Planck's work on blackbody radiation, and would overturn Newtonian mechanics at the microscopic level by showing that nature operates discontinuously (see Einstein 1961; Krips 1987; Sachs 1988; Stachel 1998; van der Waerden 1967).

By 1945, in part notwithstanding the technological advances brought about by the theory of relativity and quantum mechanics, but more correctly precisely because of them, the disciplinary structure of knowledge in which the sciences were at the privileged pole had crystalized throughout the world of knowledge; this structure would remain highly stable through the late 1960s. Institutionalized science proclaimed the Enlightenment ideal of endless progress implemented in a lawlike world, one therefore that was ultimately predictable. This science was empirical and positivist. Theoretical advances were bound up with observation and experimental verification. Science was considered to be universal, equally valid at all times and all places, and thus it defined the parameters of what knowledge could be regarded as authoritative.

In their continuing quest for legitimacy, the social science disciplines continued after 1945 to be torn between the scientism, in both theory and method, of economics (econometrics), sociology (structural-functionalism), and political science (behaviorism) and the more narrative bent of history and anthropology. Although all the disciplines exhibited to some extent both tendencies, scientism seemed to be gaining throughout, even in history and anthropology.

Similarly, the humanities, in their effort to retain a credible voice, echoed the decontextualization, atemporality, and presumptive objectivity of the sciences. The "new criticism," for instance, rebelled against the romantic exaltation of the poem as a reflection of the personality of an exceptional individual (the so-called intentional fallacy) or the impressionistic record of interior experience (the so-called affective fallacy). It tacitly privileged the individual creator and separated him (*sic!*) from the unique object, the text (conceived as a repository of meaning), to be interpreted using such approaches as "close reading."

At the very moment of the worldwide triumph of the Newtonian worldview, a deterministic world of natural laws based on time-reversible dynamics, a new knowledge movement that would challenge its premises began to take root. This movement eventually came to be known as complexity studies. For the nineteenth century had also been the century of Charles Darwin and Rudolf Clausius, that is, of evolution—of the evolution of species through natural selection, and of the evolution of the universe manifested as entropy. In contrast to time-reversibility, evolution suggested the existence of irreversible processes of real systems exhibiting a flow of time.

Complexity studies was an outgrowth of the internal advance of science itself, and as the new field developed, it addressed a topic that had been elided in the dominant Newtonian worldview: the contradiction between determinism and freedom. Paul Davies (1989) has characterized contemporary scientific research as falling into three categories—at the frontiers of the very large, the very small, and the very complex. This new appreciation of complexity, dealing with the universal features of complex systems, irrespective of the peculiar aspects of the different systems, is especially marked with regard to humanly perceivable macrosystems. It was foreseen by Warren Weaver (1948), who distinguished three varieties of problems: the simple problems of classical physics involving only a few variables, disorganized complexity with many variables and amenable to description by statistical methods, and a middle region of organized complexity in which problem-solving would depend on analyzing entire systems as organic wholes. He predicted that during the late twentieth century this latter activity—to be based on computers and interdisciplinary, "mixed team" research—would comprise the third great advance of science. He was singularly prescient.

The rethinking that we are witnessing today represents a synthetic approach as opposed to a reductionist one, strong cross-disciplinarity, and the search to include "intractable" problems (Pagels 1988; Stein 1989). Its quest for a theory of harmony in the universe "expressed by simple, but hidden, means" (Ekeland 1988: 6) marks a shift from the Newtonian worldview emphasizing equilibrium and certainty and defining causality as the consistent association of antecedent conditions and subsequent events amenable to experimental replication and hypothesis testing. Over the course of the twentieth century, the pretense of classical science to absolute space and time had been destroyed at the level of the very large by relativity, and the illusion of controllable measurements was shattered at the level of the very small by quantum theory. Only since the 1960s, however, has the last refuge of Newtonian dynamics in the non-relativistic, nonquantum domain of the human world also been challenged, if not overcome (Ford 1989: 354). By 1986, Sir James Lighthill, in his capacity as president of the International Union of Theoretical and Applied Mechanics, felt compelled to apologize on behalf of "practitioners of mechanics . . . for having misled the general educated public by spreading ideas about the determinism of systems satisfying Newton's laws of motion [implying complete predictability] that, after 1960, were to be proved incorrect" (1986: 38).

Although there is no consensus on the exact meaning of "complexity," during the last three decades it has become increasingly apparent that there exist complex phenomena that arise from very simple mechanisms (Aida et al. 1985; Atlan et al. 1985; Cowan, Pines, & Meltzer 1994; Peliti & Vulpiani 1988; Stein 1989; Waldrop 1992; Lee 1992). As Murray Gell-Mann (1994, 1995) has noted, systems with many interactions among elements but few variables may be more complex than systems that have many variables but exhibit only one-way interactions. Prigogine and his coworkers (Nicolis & Prigogine 1989) prefer to understand complexity in terms of system "behavior" rather than of system interactions. In any case, a science of complexity holds out the possibility of representing change—that is, "describing our collective reality as a process"— without reverting to reductionism (Casti 1994: 273).

The field exploded in the late 1970s, especially after the first conference devoted to the subject of "chaos" in 1977 (Gleick 1987: 184). Along with the creation of new journals (including *Complexity*, founded in 1995) and scholarly associations, PhD programs, subfields, and even new university departments, there have also been major extramural institutional ventures devoted to the study of complexity. For instance, in its website (www.santafe.edu), the Santa Fe Institute presents itself as "a private, non-profit, multidisciplinary research and education center, founded in 1984" that has sought to create "a new kind of scientific research community, pursuing emerging science" by catalyzing "new collaborative, multidisciplinary projects that break down the barriers between the traditional disciplines." Besides supporting individual scholars in residence and organizing research and study programs, the Santa Fe Institute has also become a major publisher in the area of complexity. From its inception, the aim was to create a new kind of institution "free from the drag exerted by past specialization and the tyranny of the departments" by recruiting "Odysseans," those "tortured souls" who find themselves in the middle between the Apollonians tending "to favor logic, rationality, and analysis" and the Dionysians, who "go in more for intuition, feeling, and synthesis" (Gell-Mann 1988: 14, 12).

It has generally been agreed, first, that complexity is linked to non-linearity. Second, work in the expanding field of complexity, especially of chaos, or seemingly random behavior that displays an underlying order—order-in-chaos (strange attractors); order-out-of-chaos (self-organization, dissipative structures); and visual representation of pathological functions and natural forms exhibiting noninteger dimensions (fractal geometry)—either constitutes an implicit call for a reappraisal of the assumptions of classical science or actively undertakes a reconceptualization of the objects of study, methods of analysis, and goals of inquiry long taken for granted as constituting "scientific" practice.

From the beginnings in the 1960s, work in complex systems research entailed a new kind of thinking that broke down existing disciplinary barriers. For Ralph Abraham, it was "the paradigm shift of paradigm shifts" (cited in Gleick 1987: 52). Mathematicians and physicists who had not been on good terms since the 1930s were by the late 1960s back together again (Gleick 1987: 52). From the mathematics side, differentiable dynamics, global analysis, manifolds of mappings, and differential geometry all played an important part in this new détente. However, not all developments were equally successful. From the 1960s, and especially after the publication of René Thom's *Stabilité structurelle et morphogénèse* in 1972 (English version, 1975), what came to be known as catastrophe theory, employing the mathematics of topographical theory of dynamical systems to provide qualitative models of discontinuous change in natural systems, generated wide, if short-lived, interest. Applications to physical phenomena abounded across the spectrum of the sciences but were particularly numerous in the biological and social sciences: from optics and lasers to urban growth and stock markets, from cell differentiation and ecological frontiers to aggression and prison disturbances. E. C. Zeeman, one of the principal protagonists of the movement, wrote that one of the main benefits to sociology might be

> to reinforce some of the theories of the non-mathematical sociologists. For, by providing models in which continuous causes can produce discontinuous and divergent effects, catastrophe theory may enable them to retain, indeed confirm and develop, theories which at present are being thrown into doubt by misinterpretation of quantitative data. (1977: 629)

However, a significant backlash and substantial critique of the "usefulness of CT as a tool for extra-mathematical application" developed. According to Sussmann and Zahler, it is Zeeman's theory that poses the most immediate danger. By discussing things such as prison disturbances, and actually claiming that, on the basis of their theory, they can show mathematically that certain ways of handling riots are better, the Catastrophists are creating the possibility that their advice might be followed by those who do not understand the mathematics, and are therefore unable to perceive that it is being misused. (1975: 119)

They concluded that catastrophe models exhibited a mix of questionable assumptions, faulty reasoning, wrong conclusions, ambiguous concepts, and dearth of testable predictions.

What did turn out to have enormous resonance in the long run, in terms of extensive theoretical pertinence and broad areas of application, was the study of chaos. The recognition of the existence of chaotic behavior exhibited by nonlinear systems repealed the conceit of Laplacian predictability. As these systems evolve over time, they rapidly magnify small perturbations and are thus highly sensitive to small changes in initial conditions. Despite this, there remains evidence of an embedded order underlying the seemingly random evolution of certain dynamical systems. The breakthrough came with the discovery of "strange attractors," beginning with the identification of the elegant "butterfly" attractor of Edward Lorenz. By accident, Lorenz had discovered in his rudimentary computer simulations of the weather in the early 1960s that his simple system diverged dramatically from arbitrarily close initial conditions rather than maintaining approximately the same behavior when starting from approximately the same points. The mathematical system Lorenz studied, three equations in three variables, visualized by plotting each iterative solution set graphically in phase space, neither converged to a single point or steady state nor was it a periodic loop or continual repetition. His system was infinitely complex. Nonperiodic, it never passed through any single point more than once (Lorenz 1963a, 1963b, 1964). Unaware of Lorenz's work, in 1971 David Ruelle and Floris Takens invented the term "'strange' attractor" to describe such phase-space portraits of the stable but nonperiodic behavior of a dynamical system. Tien-Yien Li and James A. Yorke first defined "chaotic" as describing a nonperiodic $\{F^n(x)\}$ sequence in 1975.

The development of chaos theory on so many fronts opened up the possibility of applying deterministic models, formerly restricted to the "closed universe" of "completely predictable systems," rather than stochastic models, to systems that behave randomly. In natural systems, not all theoretically possible states turned out to be realizable. Only some, those that lie on the strange attractor of such systems, will actually appear in nature, and here lies the limit to their predictability. This limit, however, leaves open a place for chance and therefore creativity and change, an "admirable and subtle mix of chance and necessity" (Ekeland 1998: 13, 12, 15).

Doyne Farmer and Norman Packard asserted that the "new wave science" addressed questions that "cry out for synthesis rather than reduction" (Farmer et al. 1986: viii), where research on systems involving at least two time scales is based on simulation and cuts across disciplinary lines. This was no longer a question of one problem, one answer but of following the evolution of an existing system or setting up a model, watching it evolve, and observing changes in its development as perturbations are introduced. Such simulation is computation-intensive, and paper and pencil were adequate only where linear systems were involved. None of this new work would have been possible without high-speed computing. As Ekeland has written, the computer

> reveals to the mathematician the phenomena to study and the mathematician exposes the limits of the computer. . . . The power of calculation available to men from now on is changing their universe. It is transforming their environment, it is transforming their societies, it is transforming them, it is transforming their science. Chaos theory is a beginning, not an end. (Ekeland 1998: 18, 21)

Although chaos, like complexity, with or without an emphasis on randomness, found multiple meanings (Gleick 1987: 306), a new view of the world—of the nature of physical reality, of change and predictability, and of clear disciplinary boundaries among problem sets as well as research strategies—was being born. Simple systems did not necessarily behave simply; some were found to have very complex behavior. "Hard" scientists were using mathematics in new ways and mathematicians seemed almost to be doing experiments. With Mitchell Feigenbaum's (1983) discovery of universality in cascading bifurcations of certain nonlinear systems, the mesoscale of humanly perceivable phenomena gained a universal constant that indicated that systems from such varied fields as mathematics, population biology, and fluid dynamics behaved identically in one fundamental, and measurable, way.

In the classical quest for simplicity, it was the nonlinear elements of the equations describing system evolution that had to be finessed through linear approximations; otherwise the equations could not be solved. Markus, Müller, and Nicolis (1988), however, presented the emergence of nonlinearity as a unifying principle in which universalities in a variety of open, self-organizing systems (that export entropy into their environment) offered a common language to chemists, biologists, ecologists, physicists, mathematicians, and medical doctors. Drawing on an immense body of work by Ilya Prigogine (which won him the Nobel Prize) and carried on with his colleagues in Austin and Brussels, he and Isabelle Stengers (1984) presented chaos not as the opposite but as the source and confederate of order. They considered that a conceptual transformation of science was taking place. This transformation was growing out of the challenge to Newtonian mechanics associated with contemporary research in thermodynamics focusing on nonlinearity (instability, fluctuations, order-out-of-chaos) and the irreversibility of the evolution of far-from-equilibrium, open systems, characterized by self-organizing processes and dissipative structures.

When compared with the ordered mathematical world of Newton and René Descartes, the world of Georg Cantor's sets and Giuseppe Peano's space-filling curves strained even the most fertile imaginations. Nonetheless, Bénoit Mandelbrot (1983) has shown how the structures that these (and other) late nineteenth-century mathematicians conceptualized are inherent in the everyday world around us. His fractal geometry describes shapes that do not fit into the Euclidean categories of points, lines, planes, and solids, but somewhere in between. "Twentieth-century mathematics flowered in the belief that it had transcended completely the limitations imposed by its natural origins . . . [but] the same pathological structures that the mathematicians invented to break loose from 19th-century naturalism turn out to be inherent in familiar objects all around us" (Mandelbrot 1983: 15, 3–4). Indeed, the undeniable aesthetic appeal of these self-similar structures with noninteger dimensions is due to the fact that the physical world is made up of them. Examples include Mandelbrot's now classic description of the coastline of Britain that increased in measured length as the length of the measuring tool decreased, and pulmonary and vascular systems whose volumes were found to be inadequate to hold their known contents. Even what we normally recognize as "music" (1/f, or "pink" noise) turns out to be fractal.

Fractal geometry, Mandelbrot has written, "is concerned primarily with shapes in the real space one can see," while "the theory of attractors is ultimately concerned with the temporal evolution in time of points situated in an invisible, abstract, representative space" (1983: 193). Nonetheless, fractals have also been associated with purely mathematical structures and the strange attractors representing the temporal evolution of real-world systems. The ubiquity of fractals expresses the particular combination of order and disorder characteristic of complexity studies in general and has furnished a powerful and stimulating new mode of description refocusing attention on the whole, and challenging reductionism, by recognizing scaling properties that render structures self-similar at all levels of magnification.

Implicit in all of these studies is a call for a reconceptualization of time itself, an undoing of the perspective of classical science, described as follows by Prigogine and Stengers: "The integration of the laws of motion leads to the trajectories that the particles follow. . . . The basic characteristics of trajectories are lawfulness, determinism, and reversibility. . . . The remarkable feature is that once the forces are known, any single state is sufficient to define the system completely, not only its future but also its past" (1984: 60). In the light of instability and chaos, and the association of the arrow of time with order as well as disorder, Prigogine and Driebe maintain that the laws of nature now express possibilities instead of certainties. There is no longer any contradicttion between dynamical and thermodynamical descriptions of nature. Far from being a measure of our ignorance, entropy expresses a fundamental property of the physical world, the existence of a broken time symmetry leading to a distinction between past and future that is both a universal property of the nature we observe and a prerequisite for the existence of life and consciousness (1997: 222).

The paleontologist Stephen Jay Gould made a complementary argument (1989) when he denounced the idea of evolution as either the march of progress or a cone of increasing diversity, substituting an image of diversification and decimation, history as unrepeatable and therefore unpredictable, in an exposition of the theme of contingency in the historical sciences. He also asserted the reality of irreversibility and advanced his own notion of the arrow of time: life's arrow, to replace "vague, untestable, and culturally laden notions of 'progress'" (Gould, Gilinsky, & German 1987: 1437).

Such conclusions explicitly contradict the duality of determinism and probability in our understanding of "life in the universe" expressed once again by Steven Weinberg (1994) and the duality inherent in the introduction of the so-called anthropic principle by Stephen W. Hawking along with his association of time with a geometrical cosmology (1988). But Weinberg constructs the world from the bottom up, from simple laws, and contends, echoing Ludwig Boltzman, that "thermodynamics has been explained in terms of particles and forces" (Rothman 1997: 28). For Prigogine, on the other hand, it is a top-down world:

> In order to determine whether . . . coffee is aging [cooling] I cannot consider the water molecules taken separately. If I do that I will not see the aging process. But if I consider the relationship between molecules I can then see quite clearly that the coffee is aging. We must view the encounters, the collisions and correlations between molecules in order to see the flow of time. . . . This is the time of humanity, or the time of recollections and not the time of human beings taken separately. The concept of time is dependent on a collective approach. (Prigogine in Snell & Yevtushenko 1992: 24–25)

Of course, these views remain contested. Gell-Mann contends that irreversibility might be nothing more that an artifact of "coarse graining," or incomplete information, that is, our own ignorance (1994). This is certainly not true for Roger Penrose, who would agree with Prigogine that it is the fundamental laws of physics as now understood that inhibit our understanding. Even Penrose, however, like Hawking and Weinberg, nurtures a worldview ultimately rooted in time-symmetric theories.

Prigogine sees himself as a counterforce—perhaps the counterforce—
to Weinberg, Penrose, and Hawking. . . . Weinberg [has] maintained
that philosophy had done science more harm than good [and has]
declared that, for the most part, "the results of research in physics . . .
have no legitimate implications whatever for culture or politics or
philosophy." (Rothman 1997: 29)

Prigogine could not disagree more strongly.

Robert Shaw argues that chaotic behavior is "completely ubiquitous
in the physical world." Strange attractors transmit perturbations from
the microscale to the macroscale: "The constant injection of new information into the macroscales may place severe limits on our predictive
ability, but it as well insures the constant variety and richness of our
experience" (1981: 107, 108). Prigogine, looking for the roots of time,
"became convinced that macroscopic irreversibility was the manifestation of the randomness of probabilistic processes on a microscopic scale.
But what then was the origin of this randomness?" (1996: 60). His still
controversial answer hinges on

a unified formulation of quantum theory incorporating Poincaré reso-
nances into a statistical description and deriv[ing] diffusive terms that
lie outside the range of quantum mechanics in terms of wave functions
. . . to achieve the transition from probability amplitudes to probability
proper without drawing on any nondynamical assumptions. . . . [T]he
observer no longer plays any special role. The measurement device has
to present a broken time symmetry. For these systems, there is a privileged direction of time, exactly as there is a privileged direction of
time in our perception of nature. It is the common arrow of time that
is the necessary condition of our communication with the physical
world; it is the basis of our communication with our fellow human
beings. (Prigogine 1996: 53–54)

The recognition that probability is more fundamental than trajectories
implies what Prigogine (1996) calls "the end of certainty," and the end of
certainty in scientific prediction connotes an open future of creativity
and choice. Furthermore, during a period of wide fluctuations in the
constitutive processes of a system driven far from equilibrium, including, it may be argued, a system of social relations, small fluctuations can
have enormous impact even to the extent of effecting total systemic
transformation—instabilities expanding possibilities by reducing constraints. Indeed, Ervin Laszlo (1987) has contended that human evolution is no longer genetic but sociocultural and that our age is not only
the age of uncertainty but also the age of opportunity.

The emerging reconceptualization of the natural world that is at
least beginning to be felt in the popular consciousness more closely re-
sembles our perception of the social world as unstable, complicated, and

self-organizing, a world whose present is rooted in its past but whose development is unpredictable and cannot be reversed. Not only has a significant scientific subculture of complexity studies or "complexity community" developed (Cowan, Pines, & Meltzer 1994: xvi), but also the abundance and scope of popularizations attest to the wide resonance and cultural impact of these recent developments.

Complexity studies were a direct outgrowth of developments in mathematics and the natural sciences and the social relations and structures of knowledge in which the disciplines were embedded—even including, for instance, the computing technology so often cited as fundamental to complex systems research. For the digital computer was merely an electronic, binary logic machine designed to accelerate the mechanics of the simplest computations in the service of political and economic processes in which it was embedded. Its development, in fact, was promoted by the Allied war effort in the 1940s and the drive for increased efficiency in the management of expanded business enterprises from the 1950s. The unanticipated consequences of the qualitative changes we have seen, however, suggest that speed of computation may be one of those parameters that has surpassed a threshold (analogous to a change in heat input resulting in water passing from a solid to a liquid state), thereby initiating a transformation of the relational structures of the modern world-system.

It is the symmetry of the collapse of the frontier between the humanities and the social sciences and the concurrent emphasis in complexity studies on contingency, context dependency, and multiple, overlapping temporal and spatial frameworks moving the sciences in the direction of historical social science that attests to the long-term crisis of the structures of knowledge. The vision of truth-values governed by the law of the excluded middle that has undergirded enduring ideas of classical logic, science, and common sense seems now to demand a reformulation in order to conform, in the language of complexity, to an image of possibility basins separated by fractal boundaries. Embodying an analogy between narrative and simulation, social analysts may henceforth feel licensed by the developments in complexity studies to make the shift from fabricating and verifying theories to imagining and evaluating the multiple possible consequences of interpretative accounts, and thereby reuniting "is" (the goal of science) and "ought" (the field of values) in the construction of systematic knowledge of human reality.

8

Science Studies

Norihisa Yamashita

The general rubric "science studies" is an umbrella term used here to encompass a set of both older and more recently recognized disciplines whose mix and general orientation changed dramatically over the twentieth century. The three disciplines that were constituted in the first half of the century—the philosophy of science, the history of science, and the sociology of science—shared an image of the scientist as engaged in rational and cumulative development of authoritative, truthful knowledge, in the sense of revealing how the world works. The emphasis was on the talent and effort of the individual. Those disciplines that emerged in the last quarter of the century—social studies of science, sociology of scientific knowledge, science and technology studies, cultural studies of science—took the production of "scientific" knowledge as their object of analysis and studied it from the perspective of "how science works" (Biagioli 1999: xii).

The philosophy of science, the history of science, and the sociology of science, the three older disciplines, accepted as given the same general parameters of their object of study but are readily distinguishable according to the very different analytic questions each addresses (Klemke 1988: 1–2) and the separate and distinct developmental paths each has followed.

E. D. Klemke has characterized the philosophy of science as a meta-language referring to the object-language that comprises science (1988:

6). "Whereas science is largely empirical, synthetic, and experimental, philosophy of science is largely verbal, analytic and reflective" (1988: 5). More precisely:

> Philosophy of science is the attempt (a) to understand the method, foundations, and logical structure of science and (b) to examine the relations and interfaces of science and other human concerns, institutions, and quests, by means of (c) a logical and methodological analysis both of the aims, methods, and criteria of science and of the aims, methods, and concerns of various cultural phenomena in their relations to science. (Klemke 1988: 7)

As is clear in this exposition, the philosophy of science, as practiced over the span of the twentieth century, was closely linked to the analytic tradition and especially logical positivism (see Weitz 1966 and Ayer 1959) and the Vienna Circle (*Wiener Kreis*, 1922), established by Moritz Schlick and including Rudolf Carnap, Otto Neurath, and Herbert Feigl, among others. The students of the Vienna Circle were primarily philosophers, whose concern was metaphysics and its critique.

As for the history of science, the early works were often written by retired scientists who became interested in the history of their own profession. Indeed, these writings were published in either general history journals or specialized ones dealing with the history of individual disciplines of the natural sciences, such as the *Journal of the History of Physics* or the *Bulletin of the History of Medicine*. *Isis*, founded in 1912 by George Sarton and the first real journal of the history of science, was the singular exception. Although its circulation was very limited when it started, the history of science as a field perhaps began here.

Sarton[1] was strongly influenced by nineteenth-century French thought, and especially that of Auguste Comte and his positivism. He sought to construct a single encyclopedic discipline of the history of science that would treat the development of science via the biographies of major scientists. He compiled historical facts about the lives and works of great scientists, even to an extent canonizing scientific saints. He shared Comte's views on the progressivism of science, which assumes its cumulative and rational development. Being of the generation that lived through two world wars, he was aware of the destructtive effects of science and technology. But Sarton was optimistic about the possibilities of "humanizing" the sciences by reflecting on them historically.

With the Second World War on the horizon, Sarton came to Harvard University, which had the oldest chair of the history of science in the United States. Harvard University, with Sarton, became an early mecca for science studies. Robert Merton wrote *Science, Technology, and Society in Seventeenth-Century England* here, working with Sarton, and Thomas Kuhn was invited to teach the history of science by Harvard's president, James Conant.

The origin of the sociology of science is generally linked to Merton, although he himself claims (1979, ch. 2) that its origin should be traced to Henri de Saint-Simon, Comte, and Karl Marx, which is to argue that the history of the sociology of science is as old as the history of sociology itself. In institutional terms, however, the direct origin of the sociology of science resided in the "invisible college" since the 1930s of which Merton himself was a important member (1979, ch. 1). But the sociologists did not even have their *Isis*. Most practitioners of the sociology of science were indeed trained as sociologists, except for a few professional scientists like J. D. Bernal, and most of their early writings were published in journals of general sociology.

Merton's approach is often characterized as normative, in the sense that he centers his concern on the ethos or norms of science, which he summarized in five principles: organized skepticism, disinterestedness, universalism, "communism" or communality, and neutrality. He shares with Sarton the image of the scientist as ethical hero, at least in its ideal. Of course, Merton's argument contains many other aspects that do not rely on such a heroic view of science. In any case, the two cultures gap is not an issue for him.

Karl Popper's "falsificationism" (1959; see also Lakatos & Musgrave 1970) and its normative implications have a certain affinity with the heroic/idealistic view of science shared by Sarton and Merton. For it implies that scientists "should" follow the rational procedure of positing their arguments in a form that can be "falsified," a chain of "conjecture and refutation," if the arguments are to merit the label of scientific. It is an ethical moment for a scientist, when faced with disproof, to decide to discard the hypothesis as falsified, rather than to attribute the disproof to some unessential error. To be sure, this kind of ethical moment has much less palpable reality when the practice of scientific knowledge production is routinized in a huge organized body, which became the normal operating procedure of twentieth-century science.

These analytic modes did not, however, address the transformation brought about by the emergence of "big" science, and insofar as a two cultures gap was recognized, it was considered to be bridgeable in each individual mind by means of properly designed institutional devices.

The publication of Thomas Kuhn's *Structure of Scientific Revolutions* in 1962 was to change much. Indeed, the issues addressed formed part of the ferment that was eventually expressed in the world revolution of 1968. Nonetheless, Kuhn himself did not consider his book an attack on science in any way. Nor did he promote the institutionalization of those strands of science studies that took scientific activity itself as a subject matter for analysis. His work did, however, give rise to what we might call the Kuhnian moment, a turning point that turned out to be more important than Kuhn's book or his own views of science.

In the phase of competitive expansion of the world-economy in the post-1945 period, framed as it was geopolitically by the Cold War, the sciences everywhere became more industrialized and state funded. The United States was eager to retain its hegemonic status in the field of science, an advantage that seemed to be threatened by the so-called *Sputnik* shock of 1957. U.S. state expenditure on science was immediately and constantly increased, and abundant funds trickled down around the hard core of the natural sciences, some even to science studies. In fact, the institutionalization of science studies was already taking place when Kuhn published his book, and, externally, the voices critical of received views of science were gradually becoming louder. The first major accident at a nuclear power plant took place in Canada in 1952, and the second in northern England (Windscale) in 1957. By the late 1960s there was an organized anti-nuclear-power movement (McCormick 1995, ch. 7). In 1962, Rachel Carson's *Silent Spring* was published. An instant best seller, it inspired a popular movement.

Although Kuhn was Sarton's younger colleague at Harvard, he was not a blind follower. Kuhn's work was greatly influenced by Alexandre Koyré and his "Galileo Studies" (see Kuhn 1977, ch. 2). Koyré argued that the intellectual transformation of the seventeenth century should be characterized as a parallel revolution of science and philosophy, and that the transformation of scientific theory should be placed in the context of the general transformation of the worldview of the time. Thus he established the "internal history" of science, which is the path that Kuhn followed. Quoting Willard Van Orman Quine's concept of "conceptual scheme" (Kuhn 1957, ch. 1), Kuhn underlined the dynamism of persisting in (and giving up) a conventional "conceptual scheme" with (or even without) explicit contradictions.

This is the embryonic form of what developed later to be the concept of a scientific "paradigm." Koyré not only served as "*maître*" to Kuhn but also influenced Michel Foucault via Gaston Bachelard. Kuhn's concept of "paradigm change" resonates with what French structuralists called *la coupure épistémologique* (epistemological break). While Foucault was writing the "archaeology of the human sciences," Kuhn might be said to have been writing the "archaeology of the natural sciences." *The Structure of Scientific Revolutions* pushed Kuhn rather abruptly, and even against his will, to the center of a movement reevaluating the scientific enterprise.

The book sold a million copies by 1996 and was translated into more than twenty-five languages (Sasaki 1994: 293). It became one of the most influential works of the twentieth century. Though Kuhn's thought is often discussed in terms of the concept of paradigm and/or paradigm change, Kuhn's own emphasis lay rather in the concept of "normal science," that is, the research scheme based on some past scientific achievement approved by a certain scientific community in a certain period as the basis of its members' work. The concept of paradigm is no more than this "past scientific achievement" in its core meaning. He formulates the dynamics of a scientific revolution as a series of stages: (1) the formation of a paradigm, (2) establishment of normal science, (3) normal science faced with anomalies, (4) crisis, (5) "scientific revolution," (6) formation of a new paradigm, and the cycle continues.

The theoretical thrust of this scheme hinges on the "scientific revolution" phase. Kuhn argues that there is no logical algorithm that can and should be applied to all cases of paradigm change, or transition from one normal science to another. This is what he means by the concept of "incommensurability." In this way, it denies the core assumption of logical positivism, including Popper's falsificationist version, that is, the development of science by accumulation. But Kuhn does not imply incomprehensibility or incomparability between paradigms. He just denies the common ground on which different paradigms can be evaluated at the very moment of a scientific revolution.

Thus, the book is intended to be a criticism of logical positivism, not a book of paradigm relativism, much less of antiscience. In fact, until the end of the 1960s, this book did not cause much of a popular sensation. According to Nakayama (1982), it is cited only 73 times in the 1965–1969 period, as recorded in the *Science Citation Index,* while it is cited 360 times in 1970–1974. That is to say, there was a gap between the initial rejection by logical positivists and the subsequent popular interest in its greater implications. When Kuhn was criticized by Popper and his disciples at the International Colloquium of Philosophy of Science in London in 1965 (Lakatos & Musgrave 1970), the voices of rejection of what Kuhn actually wrote were limited within science studies and more or less based on the falsificationist standpoint, whereas after 1968 there came to be a broad demand for a paradigm change on multiple social and intellectual fronts. And the idea became articulated through real antisystemic movements, such as student revolts, the environmental movement, and feminism. In these movements, the phrase "paradigm change" was virtually a synonym of "revolution"; Kuhn, against his will, was often considered an ideologue of revolution, like Marcuse (see Nakayama 1982).

Kuhn's own view of science is, by contrast, rather conservative. His normal science is meant to be literally normal, and the scientists are assumed to be faithful to the tradition except during scientific revolutions. Although Kuhn uses the concept of paradigm in many different ways, it is often associated in his writings with "textbook." He argues that scientists should undergo a period of apprenticeship with the "textbook," which paradoxically revives the issue of scientific creativity, that is, the "essential tension" (Kuhn 1977). In fact, Kuhn was bewildered and then quite irritated when he found himself misunderstood as a relativist or, worse, an antiscience revolutionary. He even attempted to withdraw the concept of paradigm after Margaret Masterman (1970) pointed out, at the 1965 colloquium mentioned above, that it is used with at least twenty-one different meanings in *The Structure of Scientific Revolutions*. In the end, Kuhn "retreated" to the field of the history of science proper and in 1978 published *Black-Body Theory and the Quantum Discontinuity, 1894–1912*. In this book, he did not use the word "paradigm" at all.

The gap between Kuhn and the Kuhnian moment, or social studies of science as an intellectual movement, which was inspired by Kuhn's works (or words, rather) cannot be understood merely in terms of the potentiality of one book or even one figure. It was a product of the critical and very special conjuncture of 1968–1973.

Indeed, that conjuncture saw the initial thrusts of a specifically critical movement. As we have seen, *The Structure of Scientific Revolutions* suggested "the fallacy of viewing scientific knowledge as absolute, objective, and universal." It "opened the door to a new epistemological understanding of science that had the potential to undermine its privileged status" (Trachtman & Perrucci 2000: 8–9). New forms of science studies with bases in the social sciences produced new (exogenous) insights into the development of science, of the process as well as of the consequences of the product. The discussions ranged from the way the social field influenced the directions of the development of science to the contingency of scientific knowledge, that is, how it was socially constructed and locally situated.[2]

Steve Fuller traces the origins of the field[3] to the political context in which trained scientists—such as Kuhn, Paul Feyerabend, and Stephen Toulmin, who did much of the early work, and their younger colleagues—were working. The pursuit of

> natural philosophy by experimental means and thereby [the acquisition of] a comprehensive understanding of reality . . . was out of place in a scaled-up, fragmented scientific enterprise that had come to be driven by military-industrial concerns. The level of disillusionment only increased among scientists who came of age in the following generation, which coincided with the Cold War and the Vietnam War. (1999: 246)

The result was a move to apply scientific method to science itself.

David Bloor's "strong program" represents the *theoretical* radicalization of post-Kuhnian social studies of science. The "four tenets" of the program ([1976] 1991: 7) are:

1. The sociology of scientific knowledge is concerned with the conditions that bring about states of knowledge. Naturally, there will be other causes than the social ones, which cooperate in bringing about belief.

2. It is impartial with respect to truth and falsity, rationality and irrationality, success and failure. Both sides of these dichotomies require explanation.
3. It is symmetrical in its style of explanation. The same types of causes explain both true and false beliefs.
4. It is reflexive. In principle its patterns of explanation are applicable to sociology itself. Like the requirement of symmetry, this is a response to the need to seek for general explanations. It is an obvious requirement of principle because otherwise sociology would be a standing refutation of its own theories.

The core of the program lies in tenets 2 and 3, namely the principles of impartiality and symmetry, which are the radicalized expression of the "principle of charity" with which Kuhn read Aristotle's *Physics*. The sociology of scientific knowledge (SSK) project then

> set out to construct an "anti-epistemology," to break down the legitimacy of the distinction between "contexts of discovery and justification," and to develop an anti-individualistic and anti-empiricist framework for the sociology of knowledge in which "social factors" counted not as contaminants but as constitutive of the very idea of scientific knowledge . . . in opposition to philosophical rationalism, foundationalism, essentialism, and, to a lesser extent, realism. (Shapin 1995: 297)

Bruno Latour with Steve Woolgar made the first major contribution that adopted the full-fledged anthropological method to observe and analyze the production process of scientific knowledge. Their *Laboratory Life* was written on the basis of Latour's participant observation at the Salk Institute for Biological Studies. Latour showed that the consequences of a radicalized form of post-Kuhnian social studies of science pointed, methodologically, to a social studies of science as a cultural anthropology of science as a culture.

Problems with dualism and the collective nature and local contingency of knowledge and its construction were emphasized in these new trends in science studies from the beginning.[4] Work characteristically crossed disciplinary boundaries to investigate the ways in which scientific knowledge is constructed (ethnographic studies, e.g., Latour & Woolgar [1979] 1986; Knorr-Cetina 1981; Lynch 1985) and how arguments are defended and findings institutionalized (including the rhetoric, politics, and power relations involved in knowledge, e.g., Shapin & Schaffer 1985; Shapin 1994). The field has now evolved to include the relationship of science to the larger world and alternative visions of what science should or could be.

The institutionalized disciplines of the philosophy of science, the history of science, and the sociology of science had been imbued with an essential trust in science itself even by those who were deeply concerned about the disastrous side effects resulting from "abuses." They were all essentially internalist. That is to say that science has its own logic whose formal objectivity distinguishes it from other forms of knowledge. Kuhn himself was not an externalist but was ready to defy radically the deep-rooted and widely shared internalism of the scientific community. However, the confrontation of Popper (and the Popperians) and Kuhn in 1965 demonstrated that it would not be easy to keep oneself detached from the internalist/externalist divide at the moment of conjunctural transformation in the light of the social dimensions of science. Despite Kuhn's insistence on the irrelevance of the dichotomy between internalism and externalism, the Popperians mounted a rigorous internalist rejection of Kuhn's argument, and multiple science-studies fronts moved toward an externalist line with straightforward social concerns.

In the 1970s and 1980s, many new academic journals reflecting the emerging concerns were launched: *Science Studies* in 1971 (which changed its name to *Social Studies of Science* in 1975), *Science, Technology, and Human Values* in 1976, *Science as Culture* in 1982, *Science in Context* in 1988. It is worth noting that the later the journal was founded, the more clearly externalist it tended to be, as the titles of the journals imply.

By and large, the externalist approach became more and more dominant and emphasized a series of "relativist" forebears—Norwood Hanson, Michael Polanyi, and Paul Feyerabend in the philosophy of science: Hanson (1958) with his concept of the "theory-ladenness" of observational fact; Polanyi (1958, 1967) with his concept of "tacit knowledge"; and Feyerabend (1975) with his radical "anarchism of knowledge." If this relativist line along the internalist/externalist dichotomy may be called the horizontal dimension, we can locate its vertical dimension in the transformation of Merton's "sociology of the scientific community" to the post-Kuhnian "sociology of scientific knowledge" within the sociology of science.

Post-Kuhnian students of the sociology of science went beyond human relations. They no longer professed a belief in the claims of science to being shielded from all social factors. They came to think that Kuhn's contribution constituted a radicalization of *Wissenssoziologie*, as it had been conducted in the 1920s by Max Scheler and Karl Mannheim, who themselves had not dared to apply their sociology of knowledge to (natural) scientific knowledge. This radicalization, or deepening of the sociological approach, was expressed in Bloor's "strong program."

At this time of the anti–Vietnam War movement, the antinuclear movement, and the denunciation of public pollution and environmental deterioration, the industrialization of science also became part of the problem, as it made it difficult for individual scientists to make a moral decision about their own research. They were surrounded by an atmosphere that pressed them to reflect upon themselves with a critical eye.

Thus, many antiscience voices were palpably linked with concrete social problems. And self-criticism by scientists was so serious that some actually changed their course of life, moving away from being scientists to being activists and/or science studies practitioners. Their condemnations of science were often total and radical, especially when compared with the early twentieth-century criticisms of scientific civilization, which had been more reformist and optimistic about the possibility of "humanizing" sciences.

However, the more dominant became externalist approaches and the more deeply researchers explored scientific practices, the more independent and the more detached from actual scientific knowledge production their field became. Postmodern approaches indicted the authoritarian and suppressive nature of modernity. The social studies of science built on new theorizing. Pierre Bourdieu's theory of the "field" (*champ*) was applied by Karen Knorr-Cetina (1983). The discursive approach was adopted by Harry Collins (1983) as well as Knorr-Cetina. Foucault's impact is evident on Nigel Gilbert and Michael Mulkey (1984). This methodological self-consciousness took on its own autonomous logic, which was no longer linked to social movements. And the target audience of the science studies movement in general, namely natural scientists, became convinced that the social studies of science had nothing to do with their activity.

Such a radical moment turned out to be a passing one. The vast majority of the scientific community had remained indifferent to the social movement against science even at its apogee. Such indifference remained the normal stance throughout the post-Kuhnian period. Academic recognition of the social studies of science was only to heighten the disciplinary wall of (more or less mutual) indifference, especially between the "two cultures." While the social studies of science was absorbed in polishing its theoretical tools—actually seldom used in the real social context—very few scientists recognized that the achievement of the social studies of science might have any practical or even indirect implications for their own activity.

In the 1990s the political implications of these theoretical trends became clearer in what might be called the "dilemma of cultural relativism" and "universality strikes back." The social studies of science succeeded in opening up the actual field of criticism (and self-criticism) of science in the period of 1968–1973, and Kuhn's contribution was to suggest that science should be treated as a culture (a product of the relativism he had to accept in his reading of Aristotle's *Physics*), in just the way anthropologists treat different cultures. Originally, this claim had a critical and leftist tonality. However, the logic backfired, and scientists, now aware that they were being treated like a cultural minority, started to assert, "You can't represent us unless you yourself are a scientist!" Malcommunication between the two cultures was thus aggressively reactivated in what came to be known as the "science wars."

Notes

1. Born in Belgium in 1884, he wrote a dissertation on Newton's statics at the University of Ghent in 1911.

2. See Barnes (1974), Bloor ([1976] 1991), Knorr-Cetina (1981), Latour & Woolgar ([1979] 1986), Shapin & Schaffer (1985); and overviews at two different times in Shapin (1982, 1995).

3. Although belying different tonalities of meaning: SSK, "sociology of scientific knowledge"; STS, "science and technology studies"; "social studies of science"; or simply "science studies." For a recent anthology, see Biagioli (1999).

4. Underlining the problematic role of the "social"—not to be conceived as "a 'dimension,' an 'influence,' or a 'factor' to be juxtaposed with the 'factors' of evidence and rationality"—Latour and Woolgar removed the term from the subtitle of the second edition (1986) of their *Laboratory Life: The Social Construction of Scientific Facts* (1979) (Shapin 1995: 300). As Shapin further observes, the claim "was that 'the social dimension' of knowledge needed to be attended to in order to understand what counts as a fact of discovery, what inferences are made from facts, what is regarded as rational or proper conduct, how objectivity is recognized, and how the credibility of claims is assessed. The target here was not at all the legitimacy of scientific knowledge but the legitimacy of individualist frameworks for interpreting scientific knowledge" (1995: 300).

※

9

The Cultural Turn in the Social Sciences and Humanities

Biray Kolluoğlu Kırlı and Deniz Yükseker

The neat divisions within the social sciences, and between the social sciences and the humanities, have now begun to be seriously questioned. The radicalism of the decolonized world, the failure of developmentalist projects, and the disillusionment with actually existing socialism have thrown into question, indeed to some extent delegitimated, both positivist and traditional Marxist formulations of social transformation. And new social movements organized around the demands and identities of underprivileged groups have forced both a rethinking of appropriate objects of intellectual inquiry and the restructuring of academic disciplines.

One of the directions taken in the quest to remedy the problems faced by social inquiry in this context was the turn to "culture" as a nexus of analysis, this often being further qualified as "popular" culture. As Zygmunt Bauman argues, whereas nineteenth-century social analysis moved to naturalize culture, a trend that reached a climax in the post-1945 period, since the 1960s we have been moving to culturalize nature (Bauman 1999: x). Of course, ordinary people had already begun to attract the attention of producers of knowledge in Europe in the second half of the eighteenth century. The mass of the population, who had been mere subjects under feudal or absolutist rule, began to be called citizens and perceived as the building blocks of nations. The romantic movement encouraged studies of folk traditions by collecting folk songs, poems, and tales in search of the true "poetry of nature" or "organic community." At the dawn of nationalism in Europe, "the wild and simple yet pure and honest characteristics" of the people became the source for the eternal truths of nations (Burke 1994: 5–8, 11).

The beginning of the twentieth century witnessed a radical turnabout. The collective behavior and culture of the masses began to acquire a negative connotation and to be treated with great skepticism. On both the right and the left, there was a disillusionment with and distaste for the common people. For the right, there was fear that the "hyperdemocratic" demands of the masses threatened the "natural order" of society. The left, on the other hand, began to fear that the proletariat was failing to fulfill its historical revolutionary role and was instead being co-opted. The Frankfurt School, for example, tended to replace the notion of the proletariat as the active agent of revolution by the notion of the passive mass.[1] When Max Horkheimer and Theodore Adorno criticized industrialized culture as escapist and taming and argued that technology undermined the very substance of art, they were talking about the colonization of high culture by mass culture, which they deemed to be immanently dangerous (Horkheimer & Adorno [1976] 1988).

In the 1950s, studies of popular culture reacquired legitimacy and since then have grown tremendously. Initially, the very distinction between high and popular culture was challenged, thereby revalorizing the cultures of the popular classes. Subsequently, under the label of cultural studies, the cultural dominance of the White, the male, the heterosexual, and the West was assailed.

In this to-and-fro of the attitude of intellectuals to ordinary people, there was a constant shift in what was meant by "the people": the people as building blocks of nations; the people as a threat to "civilization"; the people as the source of resistance to power. Since the 1950s, scholars have turned to the sphere of culture to demonstrate this possibility of resistance. We can discern four broad moments in this process, overlapping yet still with an apparent temporal sequence: (1) from the 1950s to the early 1970s, rethinking the concept of working-class cultures—early English cultural studies and "history from below"; (2) in the 1970s, formulations of hegemony and domination under the influence of structuralism; (3) since the late 1970s, emphasis on identity, subjectivity, and multiculturalism—studies of gender, ethnicity, and race; and (4) since the 1980s, post-structuralist conceptualizations of power and resistance, new historical studies of popular culture, and postcolonial studies.

The turn toward culture and the popular created an intellectual and institutional zone that was on the fringes of the established academic disciplines, and was thus labeled interdisciplinary. Here, cultural studies and the new historical studies of popular culture have developed within the humanities and the social sciences, and, we shall argue, are the outcome of a rapprochement, albeit an oscillating one, between the humanities and the social sciences.

The initial formulation of culture in its cultural studies form was in the "working-class culture" and "history from below" approach that emerged in Great Britain (or to be more careful, in England) in the late 1950s. The Communist Party Historians' Group played a significant role in the formation of this approach.[2] The group shared a concern with overcoming the reductionist base-superstructure model while retaining the centrality of class-struggle analysis in history (Kaye 1984: 5; Johnson 1979: 62). The guiding lines of their effort can be summarized as follows: First, classes were examined, not as mere reflections of the relations of production, but as historical relationships and processes. Second, the historical experiences of peasants, artisans, and workers were central to the studies of the group. They sought to recover the past that had been made by the actions and struggles of the lower classes but not written by them, hence, history from below. Their third concern was the recovery and assemblage of a "radical democratic tradition," asserting what might be called "counter-hegemonic" conceptions of liberty, equality, and community that did not originate in the thought of philosophers but were rather the products of the historical struggles of the lower classes themselves (Kaye 1984: 1–22).

Within this tradition, E. P. Thompson's work is especially important since it later became a point of reference for many in cultural studies, anthropology, sociology, and the historical studies of popular culture. In *The Making of the English Working Class,* a study of the formation of the English working class from 1790 to the early 1830s, Thompson contends that the English working class was equipped with a rich, peculiarly English tradition of democratic culture, and that in the 1790s a specific working-class consciousness began to mature with a strong democratic impulse. He argues that the formation of the working class was actually to be found in the common experiences and struggles of the lower classes—economic exploitation and political oppression. The Industrial Revolution was experienced as a catastrophe. Although industrial capitalism won, at the same time an "ethos of mutuality" and "working-class consciousness" had developed in the working-class communities (Thompson 1966: 418–44). Thus, by 1832, the working class had brought itself into being. Working people were not a passive mass but were involved in making themselves into a class. Thompson insists that a social class is not a thing, a structure, a category, or a product of mechanical determinism. Rather, it is a "historical phenomenon" and "something which in fact happens (and can be shown to have happened) in human relationships." Class consciousness is the way people handle their experiences "in cultural terms: embodied in traditions, value systems, ideas, and institutional forms" (Thompson 1966: 9–10).

The emergence of contemporary cultural studies is closely linked with the history from below project. Cultural studies in Great Britain developed around the work of Richard Hoggart, Raymond Williams (the literary strain), and E. P. Thompson (history) in the second half of the 1950s and the 1960s. Hoggart founded the Centre for Contemporary Cultural Studies (CCCS) in 1964 at the University of Birmingham. The initial phase of cultural studies was shaped within the "first" New Left of 1956. A critique of "old Marxism" and the economism that was inherent in the base-superstructure model was unleashed in the wake of Nikita Khrushchev's 1956 address to the Twentieth Party Congress of the Communist Party of the Soviet Union, denouncing the mistakes of Stalinism (Johnson 1986–1987: 39). Therefore, British cultural studies' engagement with Marxism sought to bring in topics that had previously been neglected by Marxists, even by Marx himself: culture, ideology, language, and the symbolic (S. Hall 1992: 279).

The concerns of Hoggart and Williams grew out of changes within British society. On the one hand, a mass culture was being formed through the increasing power of the media. On the other hand, the state was intervening in the educational system to instill a certain version of (high) culture in the masses. The impetus to this attempt derived in part from the recommendations of the literary critic F. R. Leavis, building upon Matthew Arnold's late nineteenth-century tradition, who argued that the "great tradition" of English literature should be taught in universities for the spiritual survival of the country (Lepenies 1988: 181). Hoggart by contrast (*The Uses of Literacy*, 1957) was sympathetic to existing working-class culture, partly created and perpetuated by the mass media of the day (Gaunt 1996: 95). The main thrust of Hoggart's work was that he was affirming "working-class culture" against the "high culture" that Leavis was seeking to promote in English studies.

Williams's approach to the dichotomy of high versus working-class culture was different from that of Hoggart. Williams endeavored in *Culture and Society* (1958) to get rid altogether of this dichotomy in culture. For instance, Leavis had claimed that "in any period it is upon a very small minority that the discerning appreciation of art and literature depends" (Leavis 1930: 3; quoted in R. Williams 1958). Williams pointed out that this argument involved an understanding of culture whereby it is perceived as a "positive body of achievements and habits" that is superior to the way of living brought about by twentieth-century "mass civilization" in the press, advertising, and films (R. Williams 1958: 254). Leavis's concern was that these institutions threatened the cultured ways that he, as Arnold before him, valued. He proposed to educate people about the literary tradition (the high culture) against the disintegrating impact of mass culture. Williams affirmed Leavis's emphasis on education, while attacking his notion of (high) culture and the idea of an elite minority, which embodies it. Williams defined culture as a whole way of life and as a "structure of feeling" rather than simply the sum of the "best that has been thought and said," as Leavis and Arnold had argued. The high/low division of culture in Leavism was rejected, and culture was now described as the giving and taking of meanings and the development of common meanings in common cultures (S. Hall 1980: 522).

The conception of culture in Williams's work challenged not only the high/low dichotomy but also the base-superstructure metaphor in Marxism. Thus culture was not just the sum of folkways and mores in a society, but more importantly, it had to do with the relationships among different social practices (S. Hall 1980: 524). By saving culture from the neglect that was inherent in the dictum "base determines superstructure," Williams accorded it a degree of autonomy. He traced how the term "culture" had acquired two different meanings in Western society since the eighteenth century: (1) the inner process, or intellectual life; and (2) the general process, or culture as the term is used in anthropology. Williams said that this dichotomization of culture had become so intrinsic to social thinking that even Marxism was reproducing the separation of culture from material and social life. Hence, Marxists could not see culture as a social process creating "ways of life." Culture as intellectual life and arts was relegated to the superstructure by Marxists (R. Williams 1977: 17–19). Thus, Marxists were committing the same error as the Leavisites: defining culture narrowly, they too selected a canon of "the best that has been thought and said" (Sparks 1996: 74).

These two critiques of culture, of Leavism and of reductionist Marxism, form the basis of the intellectual issues of early cultural studies. The work of Thompson and of Williams emphasized the importance of social agency (Grossberg 1993: 38). Hall calls this phase in the work of CCCS the "culturalist" paradigm in cultural studies (S. Hall 1980: 524).

The early work of the CCCS was characterized by an emphasis on working-class experiences. Stuart Hall started to stress as early as 1958 that the class structure of Great Britain was changing under the impact of consumerism. He argued that Williams's concept of the "whole way of life" was breaking up into a series of lifestyles (S. Hall 1958: 27; cited in Sparks 1996: 77). This point would be elaborated further in the 1970s as cultural studies moved away from the culturalist paradigm toward Althusserian structuralism after the events of 1968. Structuralism envisioned men and women as bearers of structures that accounted for their actions, rather than as active agents in the making of their own history. In other words, the culturalist paradigm had given priority to human experience whereas structuralism was arguing that experience was the consequence of the operations of the economy and ideology (Sparks 1996: 80–81). Structuralism, however, paid more attention to the internal articulation of the superstructure than to the traditional base-superstructure approach. As a result, cultural studies turned toward studies of ideology. As this shift in theoretical perspectives was taking place, Thompson (who attacked structuralism in *The Poverty of Theory and Other Essays* [1978]) and Williams withdrew from the forefront of cultural studies.

During the 1970s, CCCS produced work under the leadership of Hall (1969–1979) that placed a different emphasis on culture. On the one hand, the more humanistic bent of earlier work declined; on the other, more stress was put on hegemony and the power of the dominant discourse, a position that rejected the independent making of working-class culture. Using Gramsci's concept of hegemony, cultural studies provided a critique of culture's hegemonic effects, arguing that culture is less linked to class identity and had become more of an apparatus with a large system of domination (During 1993: 5). Antonio Gramsci's concepts were also used to analyze and demonstrate how resistance against dominant ideologies was possible (S. Hall 1980).[3] For example, they were central in the ethnographic works on youth subcultures in Great Britain (such as Hall & Jefferson 1976).

Early work at CCCS had focused on cultural forms that neglected differences such as gender, race, and ethnicity. But beginning in the late 1970s, there was a major shift of interest from issues concerning ideology and hegemony to those concerning identity and subjectivity. There was also considerable employment of Lacanian psychoanalysis in studies of ideology, which Louis Althusser had analyzed as an unconscious operation (Franklin, Lury, & Stacey 1991: 6–8), such as by the group of cultural studies scholars writing in the cinema journal *Screen*. Others, including Hall, were influenced by Ernest Laclau and Chantal Mouffe (1985). They argued that the origins of ideologies were indeterminate, and that therefore the relation between ideology and social subject was contingent, rather than determined, as Althusser had argued. Concomitantly, social class ceased to be a privileged category at CCCS. Thus, ideology could be constituted by the interests of any social group, not necessarily by those of a class. This so-called opening up of the structuralist framework was used to justify studies of such categories as gender, ethnicity, and race (Sparks 1996: 91–92). Works that illustrate this transformation in cultural studies include Stuart Hall et al., *Policing the Crisis,* 1978 (analyses of racist representations of social problems); Women's Studies Group, CCCS, *Women Take Issue,* 1978 (engagement with personal dimensions of culture within a feminist framework); Paul Gilroy, *There Ain't No Black in the Union Jack,* 1987 (constructions and representations of ethnicity and national identity); Dick Hebdige, *Subculture,* 1979 (the struggle of Black people for collective self-identity); and Stuart Hall, *The Hard Road to Renewal,* 1988 (analysis of co-optation of the working class by Margaret Thatcher's "popular authoritarianism").

The two main developments in cultural studies in the 1980s were its efflorescence in North America and Australia and the popularization of

post-structuralist and postmodern theories imported from France. Therefore, the story of cultural studies since the 1980s became more diffuse
across space as well as in subject matter. We shall mainly focus on its
trajectory in North America.

Because cultural studies has no single "address" in the United States
analogous to CCCS in Great Britain, there are different views on its
origins in the United States. Some authors point to the research done on
the impact of media industries and mass culture in the postwar period
and argue that debates about popular and mass culture since the 1960s
have formed the backbone of cultural studies in the United States (Ohmann 1991: 7). But others argue that work on culture stayed very marginal in American academia until it was "imported" from Great Britain
in the early 1980s (Spivak 1991: 68). Several factors explaining the rise of
cultural studies in North America in the past two decades have been
noted. (1) Earlier cultural studies work whose formulations were based
on the bipolarity of dominating and dominated groups (such as men
versus women, or the state and capitalists versus the working class)
seemed inadequate for the analysis of oppression and resistance at the
intersection points of racial/gender/class situations (Johnson 1991: 28).
(2) The entrance of ethnic minorities in large numbers into U.S. and
British universities in the wake of the 1960s social movements resulted
in the questioning of the curriculum (from the literary canon to the
social sciences). (3) The enormous growth in the strength of culture industries and their increasingly global reach made it seem that analyses
of media based simply on encoding-decoding (viewer-response theories)
were not sufficient to explain the impact of the media on people's daily
lives (Johnson 1986–1987). (4) European, especially French, critical and
cultural theory proved very productive in articulating new cultural concerns that emanated from these developments.

Post-structuralism and postmodernism not only challenged the via-
bility of the representation of cultures and identities, but the theoretical
tools of these approaches also made it possible to see human subjects as
articulated within multiple struggles and identities rather than being
essential subjects with an impulse to resist oppression (Grossberg 1993:
51). Deconstruction and semiotics seemed to offer the possibility of laying bare Eurocentrism in Western cultural forms, from literature and art
to social science knowledge in general. Post-structuralism and postmodernism allowed cultural studies to challenge metanarratives in general
(from Orientalism to Marxism, from sexist to racist discourses, as well as
concepts such as truth, rationality, power, and meaning).

By the late 1990s, the kinds of issues that came under investigation
within cultural studies far surpassed this scholarly field's original beginnings in Great Britain, which had focused on working-class cultures,
ideology, and hegemony, and subsequently gender and race. A classification of cultural studies compiled in 1992 included fifteen different

categories.[4] In general, as issues of difference, identity, and subjectivity have been welcomed at the front door, considerations of class and social structure have been pushed out the back door. This is apparent in recent cultural studies both in Great Britain and in the United States (Barker & Beezer 1992: 12–16).

Two important knowledge movements, multiculturalism and post-colonial studies, emerged in the 1980s. The former encompasses the studies of identity and difference; the latter grapples with issues related to the Third World.

In the wake of the civil rights and countercultural movements of the 1960s, the "monoculturalism" of U.S. society in general and of universities in particular began to be radically questioned. This was at a time when more ethnic minorities and women were entering the universities. The literary canon was now challenged as being monocultural. A growing body of scholarship insisted on the reality of cultural diversity by pointing to forms of cultural expression and production that had remained invisible or been marginalized. Multiculturalist scholarship has tackled questions of identity and difference and representations of identity, particularly the representation of the "Other." This has led to both political and theoretical problems. On the one hand, cultural reductionism involving ethnic minorities has resulted in the virtual disappearance of political economy from the analysis of the situation of marginalized groups (Goldberg 1994: 14). On the other hand, theoretical insistence on identity/difference has in effect validated the racial and ethnic categories and boundaries constructed by the mainstream. More generally, multiculturalism through the self-representation of disenfranchised groups has often led to "tokenization." Types of multiculturalist scholarship that argue for heterogeneity and hybridity rather than identity/difference have more oppositional power in that they not only debunk constructions of ethnic purity but also question disciplinary boundaries (Goldberg 1994: 26).

Whereas multiculturalism mainly addresses the predicament of the Other within core zones, postcolonial studies tackles historical and contemporary questions regarding coloniality. Although largely situated within the academia of metropolitan countries, the driving forces of postcoloniality studies are heterogeneous. One of the founding texts of what might also be called colonial discourse analysis is Edward Said's *Orientalism* (1978).[5] Said undertakes the crucial task of analyzing Orientalist discourse—with its institutions, vocabulary, scholarship, imagery, doctrines, and colonial bureaucracies—as a dynamic exchange between individual authors and the political concerns of Great Britain, France, and the United States, in whose intellectual and imaginative territory these writings came into being.

The path opened by Said was followed by both Third World and Western scholars who aimed at writing the history of these zones not as a mere reflection of European colonial domination but in terms of indigenous dynamics. Subaltern studies, the project of a group of Indian, British, and Australian scholars led for a long time by its founder, Ranajit Guha, was one the most coherent and influential projects that continued Said's line of assault on the Eurocentrist production of histories. Orientalism studied the history of "oriental" societies and cultures through language and religion, which were both perceived as frozen entities. The subaltern-studies collective tried to dismantle the Orientalist approach to Indian history, which was not only the perspective of European elites but had also been internalized by Indian nationalist elites. Influenced by the intellectual project of the history from below approach and Gramscian theory, the main pursuit of subaltern studies was to recover the politics of the people from elitist (both colonial and nationalist-bourgeois) historiographies of India (Guha 1988a: 37; Guha 1988b).

As postcolonial studies have been developed since the 1980s by scholars in Western academia, the move has been increasingly away from the initial affinity of subaltern studies with history from below and to become more closely associated with post-structuralism, postmodernism, and deconstruction—the works of Michel Foucault, Jacques Derrida, and others. For instance, Gayatri Spivak describes the work of the subaltern-studies collective as "deconstructing historiography" (Spivak 1988). Outlining the common concerns of postcolonial studies, Gyan Prakash argues that from Homi Bhabha to Spivak and the collective, the purpose of postcolonial critics is to undo the impact of Eurocentrism on the historiography of colonization. Postcolonial scholars try to achieve this, he claims, by repudiating all metanarratives and "foundational" historiographies (from the colonial to the nationalist and the Marxist) and categories (such as capitalism and the Third World).[6]

Despite the flourishing of subaltern and postcolonial studies, cultural studies per se have until now remained a largely Anglo-American-Australian academic enterprise. The periphery usually comes into consideration as the Other in relation to metropolitan "self." Or, as in the case of postcolonial studies and multiculturalism, considerations of the periphery may be co-opted into the metropolitan academic enterprise. Thus, for example, questions of race and gender are usually conceptualized for Western countries while it is left unclear if these concepts and theories would be valid for peripheral countries. More importantly, studies of culture undertaken in the core and the periphery are rarely brought together in a comparative or global perspective (Karp 1997: 288–89; Stratton & Ang 1996).

Some scholars have suggested ways of remedying the lack of a global perspective in cultural studies. For instance, Lawrence Grossberg points out that cultural studies' theoretical engagement with globalization has so far been inadequate. According to him, cultural studies will have to get rid of the "current tendency to equate culture with location in the form of identity," since this ends up "making politics entirely into a matter of representation and interpellation." Cultural studies will also have to tackle the reinsertion of spatial articulations within globalization such as emergent urban, regional, and (ethno-)national identities (1997: 10, 11). Arjun Appadurai, on the other hand, has called for a "transnational cultural studies," the basis of which should be a "cosmopolitan" ethnography, rather than the current domination of cultural studies by English literature (as a discipline) and literary studies. Like Grossberg, Appadurai is concerned with unraveling the significance of "culture" and locality in a globalized and deterritorialized world (1996: 50–52).

Ordinary people have been making their way into the academic world, particularly into historical studies, as objects of inquiry since the 1960s. Formerly forgotten groups such as women, children, bandits, poorhouse inmates, and vagabonds have increasingly become subjects of intellectual inquiry. So too have the practices of common people— witchcraft, symbolic rituals, dietary habits, and attitudes toward sexuality, the body, and death. It is now regarded as perfectly acceptable, indeed for many preferable, to study the life of a miller who lived in the sixteenth century, or the daily life of German peasants, or that of eighteenth-century apprentices in a printing shop in Paris who killed some cats (Ginzburg 1980; Sabean 1984; Darnton 1984). This kind of research, which comes under different titles, has become a serious rival to political, diplomatic, military, or intellectual history.

However, if we took this current as a whole since the 1960s, we would be hard pressed to find a consensus about the field that it constituted. Nonetheless, the new rapprochement between anthropology, literary criticism, and history gives us the clue of how to differentiate these new historical studies of popular culture from previous approaches. The works of Natalie Zemon Davis (1983), Carlo Ginzburg (1980), and Emmanuel Le Roy Ladurie (1978) share neither a choice of subject matter nor a particular intellectual heritage. What they have in common is that they promote a revival of the form of historical writing that Lawrence Stone calls "narrative." Narrative history is different from structural history in that it is primarily descriptive rather than analytical and in that it focuses on people rather than on circumstances (L. Stone 1979).[7]

Davis explains the ways in which historians can borrow from anthropology as follows: close observation of living processes of social interaction; ways of interpreting symbolic behavior; how the parts of a social system fit together; processing information from cultures that are very different. She also adds that anthropology can enhance the questions historians ask and their visions of what kinds of sources to look for in answering these (N. Davis 1981). In line with the general shift of interest from class and social and economic structures to culture, history's turn to culture displayed itself in terms of moving closer to anthropology and literary criticism.

This new perspective preferred to see the world under a microscope rather than through a telescope, to borrow Eric Hobsbawm's metaphor (Hobsbawm 1980). It drew heavily from anthropology and literary criticism instead of from economics, demography, and sociology. In doing so, it assigned a primary and autonomous role to culture. The new historians of popular culture specifically objected to formulating popular cultural forms as total structures. They argued that one cannot understand the people of the past in terms of dichotomies of true/untrue, natural/supernatural, or false/correct. Popular culture and popular rituals would not be intelligible through "explanation of any extrinsic kind" but only through a full description (Clark 1983: 75). Only by being attentive to the mood of rituals, to the particular voice adopted, or to the special gesture selected from a repertoire could the historian hope to emulate the anthropologist in capturing those elements of genre that are intrinsic to the proper understanding of rites and their emotional content (Clark 1983: 93).

In their turn to anthropology and their search for "thick description," the new historians of popular culture were faced with a challenge regarding the availability and the nature of the sources. For instance, the majority of the materials used to study popular-culture history are those that are recorded by the literate elite. We know accounts of the life, beliefs, values of the common people primarily via the writings of the literate elite, whether it be the court records; collections of folktales, songs, and proverbs; or the descriptions of festivals. In recovering the popular from the accounts of the elite there is always the danger that the accounts of the practices, beliefs, and values of the popular classes may simply be the social constructions of the culture of popular classes by the elites of the time. They could indeed form insurmountable barriers to an accurate understanding of the popular classes. Thus the debate about elite versus popular culture in the real world has been reproduced as a debate among the historians of popular culture themselves (see Muchembled 1985; Ginzburg 1980).

In a subfield of historical studies of popular culture, the history of reading, Roger Chartier has questioned the notion of popular culture by arguing that it is no longer possible to divide cultural objects or forms and social groups. "On the contrary, it is necessary to recognize the fluid circulation and shared practices that cross social boundaries" (Chartier 1989: 169). So rather than talking in terms of cultural domination and diffusion, he advises us that, "[f]ollowing Elias and Bourdieu, we can develop a means of understanding that recognizes the production of distances at the very interior of the mechanisms of imitation, competition in the midst of sharing, and the constitution of new distinctions in the very processes of disclosure" (1989: 174).

Thus far, we have tried to show that the new historical studies of popular culture, an offspring of the cultural turn, sought to bring history, anthropology, and literary studies closer. We have observed similar patterns in cultural studies. But this understanding of culture as indispensable and irreducible has resulted not only in seeing the social world as a text and everything about it as signs to read but also, and more importantly, it has shifted the form of social inquiry from explanation toward interpretation. Hence, the study of social structures, such as class, and causality in history has diminished in importance.

The cultural turn in the social sciences and history has not only had methodological and epistemological consequences. It has also had an impact on the organization of the disciplines in the universities. The pioneers of cultural studies were politically active people (some initially within the Communist Party), but they were marginal to academia. The main figures in early cultural studies—Thompson, Hoggart, Williams, and Hall—were all relegated to extramural teaching in the late 1950s and the early 1960s (S. Hall 1990: 12). So was the first institutional site of cultural studies, the Centre for Contemporary Cultural Studies. As Hall has recounted, CCCS was not welcomed by either the English or the sociology department at the University of Birmingham. Through the 1970s, most of the work at CCCS was collective in nature and came out as anthologies on subcultures, women's studies, antiracism, and education. The ur-journal of British cultural studies was *Working Papers in Cultural Studies* (1971). *History Workshop Journal* also provided a medium for publication of cultural studies work, as did the film journal *Screen*, within which Lacanian debates took shape. Centers for television and communications research in various British cities were the other institutional sites of cultural studies (G. Turner 1990: 80–84). In the late 1980s the center had to fight for survival because of university pressure to put it under the control of the English department. It became the Department of Cultural Studies but had to change its character as a graduate research center and offer undergraduate courses. Today, several universities in Great Britain have cultural studies departments or offer such programs within sociology or English departments. In Great Britain, CCCS was for a long time on the fringes of the university, but in the United States a lot of cultural studies work has come directly out of English or American studies departments. During the 1990s, a number of interdisciplinary programs on cultural studies have been established in the United States and elsewhere, and some offer graduate degrees.[8]

Outside the university, cultural studies has become a growth industry in publishing. Much of the research at CCCS had been published as anthologies, especially by Methuen and Hutchinson. Routledge, Methuen's successor, is still one of the largest publishers of cultural studies work both in the United States and in Great Britain. As cultural studies became popular in the 1980s, the number of publishing houses and university presses that carried cultural studies titles increased; likewise, journals that focus on cultural studies mushroomed during the 1980s.[9] Last but not least, academic conferences have become an important site where cultural studies research is presented and debated, and then published in edited books.[10]

A basic feature of the institutionalization of cultural studies is that it has been defined as interdisciplinary. The methods and subject matter of cultural studies derive from both the humanities and the social sciences (S. Hall 1990: 13, 16). Moreover, the institutional sites cross disciplinary boundaries. That this situation is usually celebrated reflects an antidisciplinary thrust within cultural studies scholarship. But two dangers of this process, one intellectual and the other institutional, have been pointed out. First, several authors have expressed concern about the institutionalization of pluralism and difference within cultural studies at the university (MacCabe 1992: 33; Morris 1992: 292; S. Hall 1992). Hence, once cultural critique and identity politics enter the university, they are faced with the danger of losing their political edge.

Second, and related to the first, some have argued that by being interdisciplinary, cultural studies might be articulating the new university's imperatives (Striphas 1998a: 463). Readings claims that the contemporary university endeavors to economize on resources by bringing together several disciplines under interdisciplinary programs. Thus the interdisciplinarity of cultural studies might well be serving the corporatization of the university. Indeed, "interdisciplinarity has no inherent political orientation" (Readings 1996: 39).

The institutional development of new historical studies of popular culture is not as separatist as that of cultural studies. Rather than developing around a center or a journal, proponents of such studies tried to progress by publishing historical studies that would change by their example the way in which history was done. On the one hand, the scholarship of E. P. Thompson (1966, 1978), Mikhail Bakhtin (1984), and the *Annales* school appears as some of the earlier foundation of a structure that came into being in the 1980s. On the other hand, some leading scholars within this knowledge movement were influenced by other, more contemporary, currents. For instance, the editors of the series Studies on the History of Society and Culture,[11] Victoria Bonnell and Lynn Hunt, trace their history back to the works of Hayden White, Clifford Geertz, and Michel Foucault that were published or translated into English in the 1970s (Bonnell & Hunt 1999: 2).

While cultural studies have been around since the 1960s and social history dates back to more or less the same period, the new historical studies of popular culture have discernibly proliferated only since the 1980s.[12] The leading institutional sites and periodical publications include the Popular Culture Association in the United States, the Manchester Institute for Popular Culture, the *Journal of Popular Culture*, the *Journal of American Culture*, and *American Quarterly: Popular Culture*.

The development of contemporary cultural studies and the new historical studies of popular culture coincided with, and in many ways constituted, the establishment of areas of intellectual study that challenged the disciplinary divisions of Western academia since the 1960s, such as feminism and ethnic/minority studies. This coincidence is most evident in the shift in the focus of cultural studies and the new historical studies of popular culture away from social structures and toward identities and microhistories.

These processes can be put into context within the changing structure of the world-economy in which decolonization and new social movements played a significant role in moving political and economic conflicts toward the realm of culture. More importantly, the massive entrance of previously marginalized groups (minorities and women) into the university challenged the way historical, literary, and social scientific knowledge is produced and taught. In this setting, popular culture, mass media, and cultural production by marginalized groups came under study in European and North American universities (Gripsrud 1989).

In the process of the development of cultural studies and the new historical studies of popular culture, in which a discontent with the way "culture" was handled in the traditional disciplines was critical, interaction between the humanities and the social sciences has become central. On the one hand, the humanities have started to utilize critical social scientific categories such as domination, class, hegemony, and resistance to investigate culture. On the other hand, the social sciences have begun to appropriate the epistemological concerns and the methodologies of the humanities such as ethnography, textual analysis, semiotics, deconstruction, and discourse analysis.

This is for the most part a result of the challenging of the high culture/low culture dichotomy and the Marxist metaphor of base-superstructure as well as reconceptualizations of power and resistance. As cultural practices and forms that used to be denigrated as "low culture" and only discussed in the university, if at all, under the label of folklore studies and anthropology made their way into the heart of academia, both literary studies and the social sciences were forced to find new methods and concepts to study their novel subjects. It was precisely this challenge that triggered a rapprochement between the humanities and the social sciences.

Notes

1. Not all Frankfurt School authors shared this view; for instance, Benjamin and Brecht believed in the radicalizing possibilities of a new proletarian art, a view that represented a more optimistic and positive perception of the masses.

2. Between 1946 and 1956, the group met regularly. Maurice Dobb and Dona Torr were the senior figures. The "younger historians" included Christoper Hill, Rodney Hilton, Eric Hobsbawm, Victor Kiernan, George Rudé, John Saville, Dorothy Thompson, and E. P. Thompson. The concerns of the group were reflected in the journal *Past*

and Present, beginning in 1952, as well as in several individual contributions of the members published soon after the dissolution of the group in 1956: Christopher Hill, Society and Puritanism in Pre-Revolutionary England (1964); Eric Hobsbawm, Primitive Rebels (1959); and E. P. Thompson, The Making of the English Working Class (1966).

3. It is at this point that the "second" New Left begins to have an impact on the course of cultural studies. The translation into English of Gramsci, the works of the Frankfurt School, and of Walter Benjamin by New Left Review served to open up theoretical issues about power and hegemony (S. Hall 1990: 16).

4. In an edited volume by Grossberg, Nelson, & Treichler (1992), there are sections on gender and sexuality; nationhood and nationality; colonialism and postcolonialism; race and ethnicity; popular culture and its audiences; identity politics; pedagogy; aesthetics; culture and its institutions; ethnography; discourse and textuality; science, culture, and the ecosystem; postmodern global culture; and space and time. During (1993) in his reader also includes consumption and leisure within the scope of cultural studies.

5. Said's work has enjoyed a greater popularity and influence than that of his predecessors—Frantz Fanon, Black Skin, White Masks ([1952] 1967 and The Wretched of the Earth ([1961] 1963); V. G. Kiernan, The Lords of Human Kind: European Attitudes to the Outside World in the Imperial Age (1969); and especially Anouar Abdel-Malek, Civilizations and Social Theory, vol. 1 of Social Dialectics ([1972] 1981).

6. See Prakash (1990, 1992). Arif Dirlik has argued that the concept of postcoloniality conceals the relation between the colonized and the colonizer by repudiating the axial power of capitalism. Once concepts like Third World or periphery are erased from discussion, the structural hierarchy of the capitalist world-system eludes analysis. The postcolonial situation becomes a discursive rather than a structural relationship. This has significant consequences in terms of the critique of Eurocentrism: "[W]ithout capitalism as the foundation for European power and the motive force of its globalization, Eurocentrism would have been just another ethnocentrism" (Dirlik 1999: 352).

McClintock argues that the term "postcoloniality" "effects a re-centering of global history around the single rubric of European time. Colonialism returns at the moment of its disappearance" (1992: 86). This has resulted in the classification of cultures and societies by their subordinate relation to linear, European time between precolonial, colonial, and postcolonial periods.

7. For a critique of Stone's formulations, see Hobsbawm (1980).

8. A recent compilation on the "institutional presence" of cultural studies lists nineteen universities in the United States that either have an interdisciplinary center on cultural studies or offer graduate degrees. Australia and Great Britain each have six such centers at universities. Canada has two; while the Netherlands, Hong Kong, Brazil, Austria, South Africa, and Poland reportedly have one center each affiliated with a university that deals with cultural studies (Striphas 1998b). In Turkey, too, several universities have established programs that offer degrees in cultural studies.

9. Publishers include Westview Press, University of Minnesota Press, Duke University Press, University of North Carolina Press, Verso, Columbia University Press, and Oxford University Press. A 1994 bibliographic article on cultural studies even then listed thirty-two journals focusing on cultural studies, more than half of which were published in the United States, many of which started publication after the late 1970s. Among the prominent ones are Boundary 2, Critical Inquiry, Cultural Critique, Cultural Studies, Differences: A Journal of Feminist Cultural Studies, Feminist Studies, Media Culture and Society, New Formations, Praxis International, Representations, Signs: Journal of Women in Culture and Society, and Social Text: Theory, Culture, Ideology (Kieft 1994: 1695).

10. Large conferences bringing together cultural studies people from the United States, Great Britain, Australia, Canada, and elsewhere since 1983 include "Marxism and the Interpretation of Cultures," University of Illinois at Urbana-Champaign, 1983; "Cultural Studies Now and in the Future," University of Illinois at Urbana-Champaign, 1990; "Creating Cultures," Convention of American Studies Association, 1988; "What Should Cultural Studies Be?" Modern Language Association Forum, 1988; "Relocating Cultural Studies," Ottawa, 1989; "Cultural Studies in the 1990s," University of Oklahoma, 1990; "Cultural Studies," University of Texas at Austin, 1990; "Dismantle/Fremantle," Murdoch University, Australia, 1991; "Trajectories: Towards an International Cultural Studies," Taiwan, 1992; "Postcolonial Formations: Nations, Culture, Policy," Griffith University, Australia, 1993; "Culture and Globalization," Duke University, 1994. The presentations at the University of Illinois conferences and the one in Ottawa were published as edited books (Nelson & Grossberg 1988; Grossberg, Nelson, & Treichler 1992; Blundell, Shepherd, & Taylor 1993).

11. This large series began in 1984, published by the University of California Press. The sixth book of the series, *The New Cultural History* (1989), and one published in 1999, *Beyond the Cultural Turn: New Directions in the Study of Society and Culture,* were important landmarks in terms of the self-definition of these new historians.

12. A simple library subject search of "popular culture—US" found eight entries for the 1960s, forty-one for the 1970s, and forty-six for the period 1980 to 1985. The number of books has been steadily increasing: twenty-seven books appeared between 1997 and 1998 under the same heading. On the other hand, a search for "popular culture—history—US," found no entries before 1981; for 1981–1985, nine books; 1986–1990, forty-one; 1991–1994, forty-four; and 1995–1998, sixty-seven entries. Parallel patterns are also observable for popular culture studies on France and England.

10

Gender: Feminism and Women's Studies

Volkan Aytar and Ayşe Betül Çelik

Although we can date the genesis of women's studies and a multiplicity of feminist movements and feminisms to the period following the Second World War, and in particular to the early 1960s, these developments have an important prehistory. Although their contribution is often overlooked, women have long been active in knowledge production outside of the academy. They were, for example, quite central in the creation of spaces of knowledge during the Enlightenment, as sponsors of and contributors to the salons (Anderson & Zinsser 1990).

Some scholars argue that at least since Mary Wollstonecraft's *Vindication of the Rights of Woman* (1791), there has existed a corpus of "feminist thinking" protesting and analyzing "women's social and sexual subordination and powerlessness" (M. Evans 1992: 98). Writing at a time when the economic and social position of women was in decline especially owing to increasing separation of the workplace from the family home, Wollstonecraft laid the basis for liberal feminism. Apart from her argument that women should become economically independent and socially autonomous, she criticized the reformist educational philosophy of her time, especially the one canonized by Jean-Jacques Rousseau's *Emile.* In that book, Rousseau proposed rational education in social sciences and the natural sciences for boys, while assigning music, art, fiction, and poetry to girls. Wollstonecraft argued that women were equally able to follow rational education in the realms left to the monopoly of men (Tong 1989: 14), thereby challenging the notion that women should only deal with such "emotional" or "soft" fields as humanities and art. Reforms to incorporate women and women's concerns as "subjects" of study goes back to the nineteenth century. Some have even argued that women's studies should be viewed within the framework of curricular reform beginning in the 1820s (Stimpson & Cobb 1986: 11). According to Mary Evans, during this period, "[women] wrote, participated in debates and certainly contributed to debate, but they did not, in any institutional sense, play a part in the construction of what has been optimistically referred to as higher learning" (1997: 46). Throughout the nineteenth century, the debate and the battle conducted by women in academe was mainly one of showing women's ability to match male intellectual competence; in this context, women were only allowed to function on the margins of intellectual life.

During the first part of the twentieth century, without any significant attempt to establish a separate field of study by and for women, some scattered studies presented a woman's perspective on women's lives and particular forms of gender relations. Women working in anthropology were among the pioneers. Important critical works appeared on the study of women from the 1930s through the 1950s.[1] Similar works were published in history, literature, and social sciences, with or without a theoretically grounded critique.[2] Women's increasing participation in the academy during this period did not necessarily lead to the development of an alternative ontological and epistemological perspective by women academics. A concern over asymmetric representation and the need to include women in the universities and other institutions of higher learning both as "objects of study" and as knowledge producers was often less discussed than the concern over the social role of education and educational institutions as such, as well as their relationship with power and patriarchy.[3] However, these concerns were rather limited to a critique of the social practices of the institutions instead of implying a more general critique of the ontological and epistemological precepts of the structures of knowledge.

The pioneering works of the period from the 1930s to the 1950s functioned not only to bridge what came to be known as the two "waves" of the feminist movement[4] but also to maintain "a certain level of intellectual curiosity" before the institutionalization of women's studies within the academy. But, however seminal, the power of this period declined gradually (E. DuBois et al. 1985: 17). It seems legitimate to claim that, although such women scholars as a group critically affected the coming disciplinarization of women's studies, they mainly were located in the framework of their respective fields of study. During this period, the conventional boundaries between the disciplines remained quite solid. In this context, women scholars constituted early and courageous voices raised against the claims to universality of a "male-controlled" academic world.

In the United States, the emergence of "women's studies" as a separate field seems to be indissolubly linked with the civil rights movement and the proliferation of other significant social movements. Schramm (1978) suggests that members of the women's movement, disillusioned by the sexism of both the civil rights and the peace movements during the 1960s, oriented their attention to the necessity of seeking alternative forms of both knowledge of and practice in women's lives. During the late 1960s the strain between the New Left and the feminist movement was accentuated. This development was partially to account for the creation of separate women's studies programs in the 1970s. Also the highly "value-neutral" orientation of conventional academe and its "male-centered" presuppositions[5] nourished the felt need for an "academy of her own."

The perceived necessity to produce knowledge by and for women brought about the establishment of numerous institutional sites to deal with such issues, and the early 1970s witnessed a multiplication of such formal arrangements.[6] The institutionalization of women's studies programs was dependent on funding. Private individuals and foundations, as well as departments of the federal and state governments, were among the principal initial financial sources.[7] The 1970s also saw the emergence of women's caucuses and committees within professional scholarly associations, and in some cases outside them (Howe 1991). In time, these groups became more closely coordinated at the national level, adding to the strength of the feminist scholarly movement.[8]

Outside the United States as well, feminist political movements and feminist knowledge movements worked closely together, although the particular institutional forms of feminist scholarship differed from country to country and from region to region. In much of continental Europe, women's studies programs as well as research centers outside the university structures emerged during and after the mid-1970s. A similar development occurred slightly later in Latin America and Southeast Asia. However, in these two regions, especially in the latter, a governmental presence was rather more visible (Bonder 1991; Karim 1991). Also, many research centers and programs were primarily geared toward such issues as women and development. The formation of programs and research centers lagged behind somewhat in the Near East[9] and much of Africa (Rao 1991).

Although the funding of women's studies in North America and western Europe was relatively satisfactory, at least for a brand-new field, its pace varied according to the political climate. It is not totally surprising that, in the 1980s, the neoconservative parties tended to underfund women's studies programs.[10] It seems legitimate to claim that the eagerness to fund women's studies programs was linked to the rise of the feminist movement, especially during the late 1970s.

Apart from sustaining individual women's studies programs in various universities, funding was also aimed at transforming the curricula to incorporate more material on women. The 1980s saw the emergence of vast projects of curriculum development with some of them aimed at bringing together scholarship on Black studies and women's studies.

Although in the beginning the members of women's studies programs were clear about their aims of verbalizing feminist assumptions concerning society as well as functioning as a compensatory and remedial institution (Boneparth 1978) devoted to social change in contradistinction with the "value-neutrality" of the conventional scientific disciplines and fields, some significant questions remained unanswered. One such question was whether women's studies should remain an interdisciplinary subfield or become a discipline on its own (Schramm 1978). The progressive institutionalization of individual women's studies programs failed to tame the tension between the two inclinations.

Accordingly, it may be argued that women's studies was largely shaped by three different objectives within academia: the creation of individual women's studies programs, the development of a body of scholarship that was carried out under the aegis of traditional departments, and an effort to integrate the new knowledge about women across the general university curriculum (Musil & Sales 1991: 24). Gloria Bowles and Renate Klein argue that the first two objectives implied different, almost opposite, research strategies and methodologies: "[A] feminist scholar working within one of the traditional disciplines must write to the audience in her field, she has to ground herself in the *structure and ideas set up by that discipline*" (1983: 7; italics added). On the other hand, scholars within the individual women's studies programs dealt with "that area of knowledge—women—that crosses all disciplines" (Bowles 1983: 39).

As a discipline on its own, women's studies had to produce a certain identifiable and peculiar methodology, perhaps even an epistemologically unique outlook. As an interdisciplinary activity, it could either "borrow" from other disciplinary methodologies and epistemologies or take a critical view toward them, in search of an alternative methodology/epistemology. Although "[t]he interdisciplinary circulation of disciplinary methods and materials and, more broadly, the development of various types of discipline-crossing activities" (Klein 1990: 93) had surely contributed to the comprehensive "muscle" of women's studies, this left still untouched the coherent unities and mutually separate existences of the various disciplines. Even though "it has been the goal . . . to transcend the inhibiting boundaries that divide disciplines from one another, and to achieve a fuller, more integrated approach to the study of women" (DuBois et al. 1985: 198), and women's studies had always felt the necessity to "move beyond narrow, disciplinary boundaries" and to employ "an interdisciplinary inclination" (Guy-Sheftall 1991: 310), this inclination still did not effectively challenge the very existence of disciplines and their separate and distant locations from one another.[11]

Still, a clear answer had yet to be given to the question of how women's studies could contribute to surpassing the limits of disciplinarity itself. According to Catharine Stimpson, women's studies had engaged in a process of "the construction of error," before the "reconstruction of theory" (in Schuster & Van Dyne 1985: 25). What this seemed to mean was that the critical role of women's studies, in her view, was not to generate new knowledge, or to demonstrate the weaknesses of various disciplinary paradigms, but rather to constitute an ancillary system for ensuring better disciplinary knowledge (Stanton & Stewart 1995: 2). It seemed that the division between the two cultures and between the multiple disciplines remained quite strong throughout this period and that women's studies had not yet come to grips with this issue.

Women's studies members inside and outside academe neither had homogeneous political and cultural stances nor employed identical strategies. The stance on the issue of interdisciplinarity was to a considerable degree related to the political opinions and the variety of feminism of the individual scholars. The very diagnosis of the root causes of women's subjugation had continually created differences of opinion within women's studies circles and feminist critiques in general, as liberal feminists argued that women were unjustly treated by the system, which (when reformed) would be more permissive and egalitarian.[12]

Socialist and Marxist feminists, on the other hand, claimed that liberal feminists' call for a delayed equality was strategically accurate but problematic from a wider perspective. They argued that women's oppression was linked with the workings of the capitalist system. Although Marxist and socialist feminists placed capitalism in the center of their analysis of women's subjugation, socialist feminists were also critical of the gender-blind character of orthodox Marxist thought and were flexible in borrowing from the arguments of radical and psychoanalytic feminists (Tong 1989: 173).

Radical feminists, however, diagnosed the nature of oppression rather differently than liberal, Marxist, or socialist feminists. In their analysis, women's subjugation was not solely confined to capitalism. They considered that patriarchy's legal, social, and cultural institutions were older than capitalism. Rather than a socialist revolution to liberate women, they preached a women's revolution to destroy women's ageless subordination.

In addition, the very definition of "woman" was to become a divisive issue by the 1970s. We have already argued that women's studies, as well as the women's liberation movement, owed much of its impetus to the civil rights movement and other social movements, especially of the 1960s. But the implicit alliance between the two movements would eventually break up (Hood 1984: 195). This rift had its repercussions on the academic scene too as a body of "Black women's studies" literature began to take shape.[13] The emergence of Black women's studies was conditioned to a great extent by the perceived insufficiency of women's studies to deal with the issues of women of color because it constantly set the agenda in terms of a monolithic notion of "woman." Butler suggested that "women's studies itself needs radical transformation in order to reflect all women's experience" (1989: 16).

Accordingly, the idea of "womanism," a concept coined and developed by Alice Walker and also espoused by some Third World women, was another reaction to the dominant forms of feminism and women's studies as developed mainly in the West. Walker describes a womanist as "a Black feminist or feminist of color" (Johnson-Odim 1991: 315). The whole notion may be seen as an attempt to broaden, and even to redefine, feminism and women's studies with a claim to comprehend "all" women's experiences. Chandra Mohanty called for "ethnocentric universalism," which would make feminism relevant to the struggles of Third World women. She argued that women's studies may even be charged with presupposing a false "unity of women" at a scholarly level. "The homogeneity of women as a group is produced not on the basis of biological essentials, but rather on the basis of secondary sociological and anthropological universals." What she proposed instead, was a "context-specific differentiated analysis," a tool that does not just "add" Third World women to women's studies—and feminism, for that matter—but redeploys feminist epistemology and methodology altogether (Mohanty, Russo, & Torres 1991: 55–56).

There were other powerful criticisms of mainstream women's studies, which was charged with furthering a monolithic notion of universal woman. One came from lesbian feminism, and its institutionalized version, lesbian studies.[14] Lesbian studies, although far from becoming an integral part of women's studies, secured some footholds.[15] But the challenges inspired by psychoanalysis, post-structuralism, postmodernism, and identity politics would blur the scene even further.

More and more, another internal debate seemed to occupy the agenda of women's studies—that over essentialism. Although it was longstanding, it now became central to the very nature of the scientific method (Rosser 1992) and what came to be called "identity politics." The debate around essentialism was rooted to a great degree in different perceptions coming from biology and psychoanalysis.

The 1970s and 1980s witnessed the proliferation of a series of theories asserting that human behaviors are embedded in a biological determinism (Kaplan & Rogers 1991). Especially prominent was E. O. Wilson's book (1975) asserting the key role of genetic determination. It fueled a heated debate concerning gender's possible biological roots. Some like Ruth Bleier (1984) felt that Wilson's theories were developed as a sort of counterargument to feminism. But some feminists were to champion such arguments, asserting that the concept of essential differences—biological or otherwise—between male and female need not be merely a conservative argument but could be used for feminist purposes. Some radical feminists suggested that such historically feminine traits as love, compassion, and sharing should be treasured and used as the basis of an androgynous future for human society (French 1985: passim). Mary Daly asserted that women should connect with their original, "natural self" that existed before patriarchy and release their "volcanic and tidal forces." The core, "essential . . . goodness" of women as a potentiality that may be developed through all-female experiences became an argument of some lesbian feminists (Daly 1984: 535–46). Charlotte Bunch (1986) went so far as to refuse to accept heterosexual women as real feminists. Psychoanalytic feminists like Hélène Cixous and Catherine Clément (1986), on the other hand, saw female sexuality as well as female writing (*l'écriture féminine*) as essentially different from, and more promising in their multiplicities and potentialities than, that of their male counterparts.

The critical use of psychoanalysis by feminists had already established a fertile basis for the discussion of allegedly biological differences and for the perception and cultural reading of these differences by society. Mary Jo Buhle (1998: 1) argues that the feminist and anarchist Emma Goldman was among the first to appreciate the uses of Freudianism for feminism. She claims that for Goldman's generation, feminism found a natural ally in psychoanalysis because both were striving to free the individual psyche from social bonds and limits and to bring about an "inner revolution" (Buhle 1998: 12). Indeed, although Goldman acknowledged the importance of the struggle for social and political rights, she nevertheless claimed that for genuine liberation one had to look more deeply. "The right to vote, equal civil rights are all very good demands, but true emancipation begins neither at the polls nor in courts. It begins in *woman's soul*" (Goldman n.d.: 11; italics added). She seems to have seen psychoanalysis as one of the ways to comprehend "the boundless joy and ecstasy contained in the deep emotion of the true woman" (Goldman n.d.: 6).

It may be argued that a more substantive exchange between feminism and psychoanalysis only emerged with Simone de Beauvoir, whose ideas about psychoanalysis were almost entirely negative, especially in terms of its alleged "phallocratic bias." She nonetheless saw psychoanalysis as a "tremendous advance over psychophysiology," because of its insistence on the view that "no factor becomes involved in the psychic life without having taken on human significance" ([1949] 1989: 38). In this sense, de Beauvoir saw psychoanalysis as a potentially useful tool to reject purely biological theories of gender identity and differences and to pave the way to an understanding of the "real, experienced situation . . .

as lived in by the subject" ([1949] 1989: 38). But her critique of Freud and psychoanalysis was directed to their alleged inability to account for this very situation and to their having resorted to male-centered presuppositions "by taking the male libido as their point of departure" ([1949] 1989: 49). De Beauvoir charges Freud not only with failing to comprehend the ways in which woman is constructed as the "Other" through the internalization of society's interpretation of biological differences (as opposed to actual biological differences) but also with contributing to the myth of woman as lesser than, and derivative of, man.

Feminism and psychoanalysis have always been in an uneasy yet fruitful exchange. De Beauvoir's critical interest in psychoanalysis was continued by Betty Friedan, who saw psychoanalysis as a "breakthrough" that tried to understand the repressive morality of patriarchy, and therefore as an ally of women's emancipation (1963: 92). But after acknowledging the historic significance of psychoanalysis, she claims that Freud was still a "prisoner" of his own culture and time and that most of his findings were not merely passé but also biased, and therefore not helpful in comprehending the current situation of women. "Much of what Freud described . . . was merely characteristic of certain middle-class European men and women at the end of the nineteenth century" (Friedan 1963: 94).

Another key figure of the same period, Kate Millett, also presents a negative evaluation of psychoanalysis and claims that, all in all, Freud developed a biologically determinist outlook. She nevertheless acknowledges his contributions, especially with regard to his theories of the unconscious and of infant sexuality. Her problem with psychoanalysis lay elsewhere: Freud's "entire psychology of women, from which all modern psychology and psychoanalysis derives heavily, is built upon an *original tragic experience—born female*" (Millett 1971: 178; italics added). She claims that Freud fails to provide any explanation whatsoever why girls would be dramatically influenced by the fact that they "lack a penis" without resorting to the presupposition that "maleness is indeed an inherently superior phenomenon" (1971: 178). Millett argues that Freud, by preferring a facile "biological" explanation, missed an excellent opportunity to study the real roots of the male supremacist culture.

During the same period in which Friedan and Millett, among others, were developing their dismissive accounts of psychoanalysis, Juliet Mitchell's work (1974) reintroduced psychoanalysis in a more positive sense. She basically charged all previous feminists with misunderstanding Freudian theories as biologically driven and failing to see their explanatory power. "[W]hat Freud did was to give up [biological explanations] precisely because psychoanalysis has nothing to do with biology—except in the sense that our mental life also reflects, in a *transformed way, what culture has already done with our biological needs and constitutions*" (Mitchell 1974: 401; italics added). Mitchell sees psychoanalysis as a potent theory to explain this very transformation. She particularly values Freud's discovery of the unconscious, understanding of which "amounts to a start in understanding how ideology functions, how we acquire and live the ideas and laws within which we must exist" (1974: 403). According to Mitchell, then, psychoanalysis is a comprehensive framework to illuminate the "unconscious structure" of patriarchy.

Some scholars argue that Freudianism was never accepted in its orthodox formulation in feminist circles, and that women's studies and feminist scholars conceived a need for reformulating his theories (Buhle 1998). Two such "rereadings" of psychoanalysis are object relations theory, which was mainly developed in the United Kingdom and North America, and Lacan-inspired feminist psychoanalysis, which has flourished in France and other parts of continental Europe.

Object relations theory, which concentrates on the pre-oedipal relation between mother and child, has led "to a normative emphasis on motherhood" (Wright 1992: xvi). Although object relations theory later became part of the mainstream within psychoanalysis, one can claim that, at its beginnings, it paved the way to its critical reexamination. Nancy Chodorow, indeed, emphasizes the (ignored) importance of "mothering" in the sexual division of labor in societies and the need to analyze "the reproduction of mothering as a central and constituting element in the social organization and reproduction of gender" (1978: 7). She claims that psychoanalytic preoccupation with libidinal development accounts for its failure to explain the centrality and importance of mothering and identities shaped by this act. By viewing mothering and mother-child interaction through an object relations lens, one could, according to Chodorow, comprehend how and why a "relational identity" develops in women, when men come to attain a "positional identity." Object relations theory, then, proposed to explain the mechanisms of reproduction of gender and gender identities, and therefore of patriarchy, through the centrality of motherhood.

Lacan-inspired feminist psychoanalysis, on the other hand, sought to develop one of the features of his theory that seemed "most relevant for feminism," that is, "[Jacques Lacan's] formalization of a logical subject, not a biological one" (Ragland-Sullivan 1992: 205). Feminists closer to that position argued that Lacan developed Freud's "non-biologist thinking to its logical conclusion by showing how men and women are *constructed* in a patriarchal system" (Wright 1992: xvii; italics added). Following this line of thought, Ragland-Sullivan (1982) argues that Lacan avoided the pitfalls of biological essentialism and tried to show that man and woman were made in culture, not created in nature. Luce Irigaray, on the other hand, by trying to undercut the mastery of phallocentric discourse, asserts women's specificity. According to her, women's erogenous "multiplicity" emphasized her biological difference (Irigaray 1985: 30–33).

Buhle suggests a cautionary note on the commonality of feminism and psychoanalysis. She claims that in their critical reexamination of Freudian orthodoxy, neo-Freudians and feminists "sharpened the significance of two opposing points of reference in both feminism and psychoanalytic theory: biology and culture, nature and nurture. "The chapter they opened has yet to be closed" (1998: 10). Through their long exchange and dialogue, feminism and psychoanalysis may be said to have asked very important and critical questions that have challenged the separateness of these "reference points," but this exchange and dialogue seem to have produced ambiguous results in terms of the construction of alternative reference points beyond this dichotomy.

As one of its central concepts—that of universally oppressed women—started to erode and critiques of its presuppositions emerged, women's studies moved to new emphases on studies of women of color and Third World women, lesbian feminism, and the importance of psychoanalytic concepts. The epistemological and methodological repercussions of this erosion implied a reconceptualization of the basic notions of what is to be known and who can know. Some of the more recent critics of earlier women's studies argue that a homogenizing notion of a "universal woman" hampered rather than helped the development of an alternative structure of knowledge from within women's studies. The charge against the notion of a universal woman appears to be not merely a political one but also an epistemological one. As such a presupposed universality entails ahistoricity and acontextuality, Mohanty suggests a "context-specific differentiated analysis" as a fruitful alternative to the limits of dominant forms of feminist epistemology (1991: 67).

The acknowledgment of women's "diversity"—understood both as a "differentiation" in women's experience, appearing as a "dissimilarity," and as a "variation" (de Groot & Maynard 1993: 150–51)—implied that the loss of the universal woman may in fact be seen as a promise of greater richness. Women's studies, it was argued, was rediscovering—or was being faced with—the plurality of explanations, the multicausality of phenomena, the diversity of social contexts, and the historicity of contingencies. Facing such multiple situations and trying to make sense of them may have, in fact, provided women's studies with a new epistemological device more comprehensive than dominant ones. But such a possibility can be fulfilled only after one locates the kinds of epistemological newness that women's studies claims to bring about in terms of a radical transformation of the structures of knowledge. In order to evaluate this, one has to look at the feminist critique of the sciences.

Feminist critiques of knowledge systems were especially successful in the social sciences and the humanities, in terms of the degree to which they affected and were taken seriously by the disciplines and their hegemonic epistemologies and ontologies. In contrast, the impact on the natural sciences has been modest, largely because natural scientists' insistence that science consists of disinterested, objective, and proven knowledge has made it very difficult for feminist scientists directly to challenge this belief (Grosz & Lepervanche 1988: 5).

The earlier feminist critiques of science had mainly been concentrated on the question of the scarcity of female scientists. Feminists charged that this feminine "absence" led to the erosion or distortion of things seen as "female" by science, which was accused of being a predominantly "masculinist" endeavor. The critique limited itself to showing the "male bias" in science's presuppositions and methodologies to pointing to "women worthies" in science's past. But this early feminist stance left the basic premises of science untouched since it implied that, when conducted "accurately" (without "male bias" and with more emphasis on "women's contribution"), the scientific endeavor would be able to carry out its noble mission of unmasking the secrets of nature. This effort had been termed "feminist empiricism," defined as the argument that "sexism and androcentrism are social biases *correctable* by stricter adherence to the existing methodological norms of scientific inquiry" (Harding 1986: 24; italics added). The shift from a "women question in science" to the "science question in feminism" implied a more radical critique of science, of both its premises and presuppositions on the one hand and its methodologies and human consequences on the other. As Sandra Harding suggested for the case of biology, "it turned out to be impossible to 'add women' without challenging the foundation of these disciplines" (1991: 47).

Ecofeminism, for example, viewed current scientific activity, together with modern technologies, as basing itself on false dichotomies and producing disastrous human and natural costs. "Science's whole paradigm is characteristically patriarchal, anti-nature and colonial and aims to dispossess women of their generative capacity as it does the productive capacities of nature" (Mies & Shiva 1993: 16). Ecofeminism also argued against basic epistemological and ontological notions of science: "The epistemological assumptions . . . are related to ontological assumptions: uniformity permits knowledge of parts of a system to stand for knowledge of the whole" (Mies & Shiva 1993: 24). Noting women scientists' critique of science for its "context-stripping" (Hubbard 1988: 10) features, Maria Mies and Vandana Shiva claim that those assumptions bring about "context-free abstraction of knowledge" (1993: 24), which represents itself as an "objective" stance. Together with these critics, ecofeminists propose instead an alternative feminist research strategy. It would include such notions as "conscious partiality" as against conventional science's claim to being "value-free"; breaking down of the vertical barriers between the researchers and "research objects"; and making of the research process itself a process of "conscientization" (Mies & Shiva 1993: 38–41).

The ecofeminist critique of science, like their proposed alternatives to it, has some similarities with that coming from radical feminists working within the natural sciences. The most general critique has been directed toward science's "masculinist" nature, its insistence upon the distance between the object and the subject, and its claim to be value-free and progressive (in terms of being beneficial to humanity). As an alternative, feminist methodology proposed to recognize the "'indisputable unity' between subject and object" (Hubbard 1988: 49). The alternative model to the standard scientific split between the knowing subject and the knowable object was derived from within women's own experience. One scholar's alternative included an acknowledgment of the unity of the knower, the world to be known, and the processes of coming to know, dubbing this the "unity of hand, brain, and heart, (an) activity characteristic of women's work" (Rose 1983, cited in Harding 1986: 142). Although such approaches run the risk of giving priority to—even essentializing—women's experience, they provide useful hints. In this regard, the historical search for a gynocentric science[16] and the appraisal of women's health movement as a "concrete example of a different kind of science" (Hubbard 1989: 14) both indicate how the feminist critique has been able to bring about rather different contextual "readings" of the history of science, unlike totalizing narratives that have left out other historical alternatives. This proposition is not only a challenge to science, forcing it to come to terms with its gendered character, but also a radical critique of its epistemological premises. "The paradigmatic challenge of feminist research is to develop a scientific institution that views gender as not only a social category, but also a category of the theory of science" (Saarinen 1988: 47). This criticism tries to show that science is historically embedded in asymmetrical power relations within a society built on rifts of race, class, and gender. Donna Haraway suggests that not only the human sciences but also the natural sciences are "culturally and historically specific, modified (and) involved" (1989: 12), a contention that strips science of its objectivist and value-neutral garb.

The feminist critique of science has not been a solely dismissive one, rejecting all analytical categories and pointing to the impossibility of any kind of knowing. Indeed, Haraway specifically warns against the dangers of falling into the pitfall of "epistemological anarchism" while criticizing science. "An epistemology that justifies not taking a stand on the nature of things is of little use" (1981: 480). She seeks to show that science, far from being neutral and objective, is embedded in the asymmetries in society. The argument that science presupposes false dichotomies is demystifying, but it is not dismissive. When Evelyn Fox Keller suggested that history of science offers the "thematic pluralism" necessary for the development of a "new consciousness" (1982: 601–2), she insisted nonetheless on her faithfulness to a living ideal of knowing the world. Indeed, the feminist critique claims to be more faithful to this ideal than conventional science, with its "pose of disinterested objectivity [that makes] 'concrete objectivity' impossible" (Haraway 1989: 13).

With the shifting of focus of women's studies and feminist critiques from attempts to increase female representation and the introduction of more women's voices in academia to more general questions of disciplinarity and epistemology, various disciplines faced problems with which they had great difficulty contending. Those problems were not limited to the "question of women" but included the introduction of new phenomena that seemed incomprehensible when divided into separate compartments. This served as a reminder of the complexity of the social and natural worlds and of the fact that the separate disciplines seemed unprepared to deal with it, owing to separateness and their habit of "simplifying" problems. In that sense, women's studies and feminist critiques sought to force individual disciplines to acknowledge their limits, a problem that was not easily resolved by the mere assertion of the virtues of multidisciplinarity. This situation was compounded, as Susan Sheridan (1991: 45–47) warned, by the dangers of becoming too attracted to the academy by its material benefits, such as tenure, accreditation, and having a budget, which could eventually bring about the absorption of the women's studies movement.

In addition to the questions concerning disciplinarity, women's studies and feminist critiques provided a less direct but still meaningful challenge to the separation among the three supradisciplinary categories: natural sciences, social sciences, and humanities. This challenge was less direct in the sense of being less frequently stated and of producing less practical results, but it was still meaningful in the sense that it has presented hints that, when combined with contributions of other alternative knowledge movements, may work toward surpassing the legacy of the two cultures. As Jane Flax suggests, to understand the promise and limitations of feminist theories, one should locate them "within the wider experiential and philosophical contexts of which they are both part *and* a critique." She claims that feminist theorists "need to enter into dialogues with "other(s)" (Flax 1990: 27; italics added), which seems to mean other alternative knowledge movements.

By forcing scientists to ask "if scientific claims to knowledge are any better than social scientific or humanistic claims" (Keller 1989: 34), feminist critiques have placed science on the same truth plane as the other two domains. Science had drawn its claims to superiority from a set of justificatory strategies, including its "nonembeddedness" and its ability to produce certainties. Feminist critiques have sought not only to show that science itself and its epistemological tools are socially and historically specific and embedded but also to demonstrate that these certainties were fragile. Far from being dismissive, feminist critiques have also sought to initiate a transformation of the logic of science, one that will "decisively break with assumptions about the autonomy of the natural sciences" (Harding 1991: 308). By rejecting the dichotomies of subject/object, self/other, reason/passion, and mind/body, feminist critiques have also sought to contribute to a rapprochement between science and the humanities by forcing both to leave their confines protected by the very existence of these dichotomies and by helping to open up a possible relational middle ground.

Women's studies and feminist scholars had to make their way in a hostile environment, one not at all eager to listen to what they had to say. Their initial theoretical and methodological fuzziness stemmed from the problems of ascertaining exactly what needed to be criticized, dismissed, and reconstructed. Their focus slowly shifted from an effort to add women—to the structures of knowledge, to scholarship, and to the academy, in the many senses of adding—to an effort to problematize the very structures of knowledge and their scholarly and institutional manifestations.

Notes

1. Among the best known of such studies were Mead (1935, 1953), Landes (1957), Leith-Ross (1939), and Kaberry (1952). These books not only vividly documented the life experiences of women from various cultures, but also in Mead's case, for example, investigated the "plasticity" of gender roles (E. DuBois et al. 1985: 16).

2. Mary Beard tried to explore the ways in which women were a "force in history" (1946). Eleanor Flexner, in her cornerstone work, *A Century of Struggle* (1959), attempted to present the history of feminism itself as a subject of study. Virginia Woolf, *A Room of One's Own* (1929), and Simone de Beauvoir, *The Second Sex* ([1949] 1989), were seminal, even epoch-making, works in their respective fields.

3. It is interesting to note, for example, Virginia Woolf's shifting emphases in her two works *A Room of One's Own* (1929) and *Three Guineas* (1938). Mary Evans argues that Woolf's "tinge of envy" toward the university in her first book is replaced by a far more critical tone in *Three Guineas*; "[T]he universities are in part about maintaining the power of the military, the established church and paternalistic figures of authority in the culture and community" (1997: 49). Woolf's shift took place within a climate of increasing criticism about the relationship between power and knowledge. Robert Lynd had launched his powerful attack on the universities of the United States in *Knowledge for What?* (1940), and in Germany, members of the Frankfurt School had raised questions about the nature of intellectual and academic work and its social functions (M. Evans 1997: 50).

4. The first wave of feminism refers to women's efforts to attain and broaden their access to political power from the late nineteenth through the first half of the twentieth century. Second-wave feminism is usually considered to have started with the publication of Betty Friedan's *The Feminine Mystique* (1963). J. Evans argues that second-wave feminism's early stage, which was a liberal one, was born out of Friedan's work and the U.S. presidential commission of the 1960s, whereas its later stage, namely the radical one, came out of the 1960s New Left, and the movement for Black civil rights (1995: 13).

5. In history, feminist critique argued that "historians' analytical concepts and frameworks have taken the male experience as the norm for humanity" (E. DuBois et al. 1985: 19). Joan Kelly-Gadol (1976), for example, illustrated the case in historical categories of analysis and periodization. Social history seemed to be more open to the

women's voice, as it had set itself the goal of rediscovering the private life, a long neglected component of history. E. DuBois et al. (1985: 20) argue that the growth of the new social history approach may be linked to the context of the political and cultural liberalism of the 1960s. In anthropology, the critiques of "male bias" were raised by such scholars as Sally Scolum (1975) and Michelle Z. Rosaldo, Louise Lamphere, and Jean Bamberger (1974). It is striking to observe the similarity between the charges of "ethnocentrism" and "androcentrism" within anthropology. In education, the issue of sex discrimination was raised by such scholars as Myra Sadker and Nancy Frazier (1973), among others.

6. The first institutional approval for a women's studies program came in 1969 at San Diego State University (Musil & Sales 1991: 23). The Women's History Research Center at Berkeley, Cornell University's female studies program, and the women's history program at the State University of New York at Binghamton were among the pioneers. A sixfold increase in the number of courses about women in 1971 when compared with the preceding year illustrates the trend. By 1976, there were 151 formal programs on women's issues. "By 1977, when the National Women's Studies Association was founded, there were 276 formal programs. By 1980, the number had increased to 332; in 1989, there were 530. In 1986, there were 334 undergraduate concentrations, certificates, or minors. In 1989, that number jumped to 404. At the graduate level, the numbers increased from 23 to 55 graduate certificates, concentrations or minors" (Musil & Sales 1991: 27).

7. The Fund for the Improvement of Postsecondary Education (FIPSE) was among the earliest to fund women's studies in the 1970s. Federal institutions funding women's studies included the National Institute of Education (NIE) and the Women's Educational Equity Act Program (WEEA). Among the earliest private funders were the Ford Foundation, the Carnegie Corporation, the Rockefeller Foundation, the Rockefeller Brothers Fund, the Andrew W. Mellon Foundation, the Helena Rubenstein Foundation, the Russell Sage Foundation, the Exxon Education Foundation, the Eli P. Lilly Foundation, and the Revson Foundation (Musil & Sales 1991: 23).

8. A central event in the institutionalization of feminist scholarly movement in the United States was the founding of the National Women's Studies Association in 1977. Howe (1991) argues that this development, coupled with the establishment of national fellowship programs for research on women, and the establishment of research centers on women contributed considerably to the institutionalization of women's studies.

9. However, Lebanon was a significant exception. The Institute for Women's Studies in the Arab World came into being there in 1973.

10. Rosemary Auchmuty (1996) illustrates the case for the United Kingdom, where, under successive Conservative Party governments, women's studies and lesbian studies programs came under fierce financial attacks.

11. Meeth (1978: 10) distinguishes between what is "cross-disciplinary, viewing one discipline from the perspective of another (art history is an example); multidisciplinary, presenting the way a number of different disciplines view a single problem; interdisciplinary, which suggests an integration of disciplinary perspectives; and transdisciplinary, beyond the disciplines." Bowles (1983: 40) suggests that transdisciplinarity should be the goal to be attained by women's studies.

12. An overview of liberal feminist arguments may be found in Jaggar and Rothenberg (1984). See especially Joyce Trebilcot's piece, which employs a liberal feminist perspective.

13. Black feminist scholars, such as Phyllis Wheathley, Anna J. Cooper, and Zora N. Hurston developed important theoretical accounts of Black women's oppression

by racist power structures in society (Guy-Sheftall 1991: 306). Some of the influential publications included Bambara (1970), Lerner (1972), Hull, Scott, and Smith (1982), hooks (1982). Guy-Sheftall notes that "the founding of *SAGE: A Scholarly Journal on Black Women* in 1983 was a milestone in promoting research throughout the world and an obvious manifestation of the 'coming of age' of Black women's studies" (1991: 306).

14. Auchmuty argues that although postmodernism claims credit for the introduction of the notion of "difference," "it was in fact the radical feminists of the mid-1970s who made it possible for lesbian and also black, working-class and other voices to be heard within feminism and women's studies, by pointing out that they had been excluded from analyses of white, middle-class, heterosexual women and that their situation, in some respects similar, was in other respects quite different" (1996: 202). This view also "problematizes feminist 'standpoints'—that either marginalize or misrepresent the experiences" of the working class, lesbian women, and women of color—themselves (Hallam & Marshall 1993: 76).

15. Stimpson notes that in the United States, lesbian caucuses were formed within the academic disciplines and, in 1977, within the National Women's Studies Association (1992: 377–78).

16. For example, Ginzberg (1989: 72) claims that gynocentric science has always existed, arguing that the "arts" of midwifery, cooking, and homemaking were activities that, if practiced by men, "would have been labeled as science."

11

Regional Categories of Analysis: Latin/o Americanisms

Agustín Lao-Montes

Regional categories of knowledge production and analysis have a long history. We are taking the case of Latin America as such a category as an instance of the more general process and shall particularly concentrate on the development of the multiple Latinamericanisms produced in the United States since the era of U.S. world hegemony after 1945. There are three different meanings of the term "Latinamericanism." The most common usage is Latin American studies, a variety of area studies. In this usage, there is not much problematization of the term. Latinamericanism can also refer to a type of imperial/colonial discourse (analogous to Edward Said's Orientalism) that produces an object (Latin America and Latin Americans) and a particular type of authorized intellectual subject (the Latinamericanist). There is a third meaning, which places itself in opposition to the second, Latinamericanism as a variant of postcolonial theory, which engages in a critical analysis of the coloniality of modern regimes of power and knowledge from the standpoint of Latino/Latin American subalternities.

Of course, the critique of Western structures of knowledge from Latin American perspectives, as well as more general challenges to the racialized character of Occidentalist knowledge, is older than this. For instance, Latinamericanism as a discourse of regional self-definition (Ramos 1989) was produced by Creole intellectuals and ruling classes in the late nineteenth century to differentiate Latin America as an emerging world-historical region from its external others (Europe and the United States) as well as from the internal others (Blacks, Indians, Chinese) of the nascent nation-states. In his analysis of the 1898 Spanish-Cuban-American-Filipino–Puerto Rican war, Uruguayan intellectual José Enrique Rodo (1988) defines the difference between the United States and Latin America as one between Anglo-Saxon scientific pragmatism and Hispanic-American romantic aestheticism. This civilizational binary, which corresponds to C. P. Snow's division between the two cultures (1959) (the sciences and the humanities), revealed and codified enduring tendencies in the self-definitions of the two Americas. It informed not only ideologies of identification but also the organization and institutionalization of the production of academic knowledge. This hemispheric civilizational/racial divide (the Anglo-Saxon race and the Hispanic race) demarcated two geohistorical fields of knowledge production that would not fully meet until the creation of Latin American area studies in the 1940s.

From another American angle, perhaps the most important historical challenge from the standpoint of "race" and/or "ethnicity" to the modern structures of knowledge came from a long-standing tradition of African-American intellectual culture (Hanchard 1990). In the early twentieth century, U.S. Black intellectual W. E. B. Du Bois, a pioneer of historical sociology in the United States, developed an intellectual, political, and pedagogical project in which he analyzed the central role of race in modern history, advocated the decolonization of intellectual production from the legacies of slavery and structural racism, and promoted the democratization of Western (and in particular U.S.) institutions, partially by promoting critical thinking through humanistic education as a means to racial democracy (Du Bois 1968 [1940]). In the same vein, the Martinican poet, essayist, and activist Aimé Césaire launched in the 1940s a critique of Western modernity from the perspective of the colonized and racialized others of imperialist capitalism, becoming a leading theoretician of anticolonial movements precisely at the time of the inception of area studies (Césaire [1955] 1972).

Du Bois and Césaire shared a critical posture against the colonial undersides of capitalist modernity, and consequently both of them denounced the dehumanization of the racialized subalterns of Anglo-European domination, those subjugated by slavery and overexploitation, exclusion from citizenship, devaluation of self and memory, and deprivation of wealth, power, and recognition. But they also shared an identification with the democratic and emancipatory project of modernity. Indeed, both Du Bois and Césaire identified with the project of human progress and the liberation of Western democratic discourse and its structures of knowledge. They stood for the democratization of world power relations (political, economic, cultural), in which the full enfranchisement and empowerment of colonized/racialized subjects was seen as a crucial problem and an absolute test of modern progressivism. Later postmodern and postcolonial critics would define the Western episteme as being based, in principle, on "racialized knowledge" (Goldberg 1993; Mills 1997; Young 1990). In short, Du Bois and Césaire were key figures in an Afro-diasporic tradition that challenged in various ways (in both theory and practice) the structures of power, the structures of knowledge, and the structures of feeling of Western modernity. This critical Afro-diasporic literature about the modern, which can also be analyzed as countermodern (Gilroy 1993; Patterson & Kelley 2000), constituted a specific locus of production and circulation of knowledge, one that was marginalized and/or erased from the dominant structures of Western knowledge. In the case of the United States, in spite of the existence of Black universities, research centers, scholarly groups, and publications since the early twentieth century, it was not until the demands of cultural justice and academic democracy of the antisystemic movements of the 1960s and 1970s that African-American challenges to the structures of knowledge, from the particular perspectives of African-Americans as racial subordinates, gained prominence and visibility.

In contrast, in Latinamericanist discourses the question of race had historically been overridden by two entwined all-encompassing identities: world-region and nation. In general, Latinamericanist state nationalisms have been sustained by an ideology of racial democracy, in which the principle of miscegenation (*mestizaje, mulataje*) has served to deny racial difference in favor of national unity, while at the same time combining Western civilization with racial hybridity as a road to "whitening" the nation (Radcliffe & Westwood 1996). This peripheral variety of Occidentalism has been constitutive of Latin America as a world-historical region (and its nations) since their inception in the nineteenth century, deploying some of the main categories and antinomies of Western thought (civilization/barbarism, culture/nature, modernity/tradition), to distinguish Creole ruling classes from racialized/ethnicized (as well as gendered/eroticized) subalterns.

Even most of the Latin American Marxisms that emerged in the 1920s participated in the same putative denial of race and racism, in favor of Western-looking progressivist ideologies advocating socialist cosmopolitanism and/or revolutionary nationalism (Arico 1980; Löwy 1982). A highly influential exception was Peruvian Marxist theoretician and revolutionary activist José Carlos Mariátegui, whose theorizing and politics were grounded on an analysis of the specificity of Andean socioeconomic structures, political regimes, and the corresponding class and ethnic relations (Mariátegui [1928] 1971). However, not even his proposal for an Indoamerican socialism took account of the centrality of the racial problem, and particularly of the place and role of people of African descent in Latin American/Caribbean histories of power, resistance, and transculturation.[1]

Thus, the prominence of race as a category in Latinamericanism occurs only as of the late 1960s and may be said to be due to the impact of the antisystemic movements. The issue of the "liberation of subjugated knowledges"—those of Blacks, Indians, women, the marginal poor (Foucault 1977)—does not arise until the 1980s (Yúdice, Franco, & Flores 1992). In these developments, the histories of Latin American area studies and of Latino ethnic/racial studies are intertwined.

The period after 1945 was marked both by the establishment of the United States as a hegemonic power in the capitalist world-economy and by a widespread wave of national liberation struggles for decolonization. The building of world hegemony involved, in addition to the creation of systemic frameworks for capital accumulation and interstate institutions (like the United Nations and NATO), an ideological terrain that would facilitate U.S. (and Western) intellectual and moral leadership. This drive to what we may call cultural and epistemic hegemony required a substantial growth in the U.S. university system, a globalization of Western academic conventions and institutional forms, a consolidation of the ideology of the two cultures, and the reinforcement of disciplinary boundaries by means of specialization and professionalization (Lee 1996; Manicas 1987; Gulbenkian Commission 1996).

Area studies played an important role in the new situation. It emerged primarily within the social sciences, with the manifest geopolitical objective of producing useful research on world regions (like Latin America and Asia) that were seen both as zones of "development" and zones of turbulence, given the movements for decolonization, especially during a period of cold war. But the project of area studies also involved a will to globalize Western values, knowledge, and institutions by means of a new version of the civilizing mission that was embodied in modernization (developmentalist) theories. Area studies was generated by the concerns of governments (particularly the U.S. government) and of large foundations (like Ford and Rockefeller), which pursued their solutions largely via the research universities, once again particularly in the United States. Latin America was a priority region from the inception of area studies, setting the stage for large-scale, well-financed, U.S.-based Latin-americanism.

Area studies were conceived as an interdisciplinary field to be practiced by teams of researchers mostly from the social sciences (political scientists, historians, economists, anthropologists) who were expected to maintain their primary loyalty to their disciplines. But the disciplinary knowledge (in the double Foucauldian sense of disciplined and disciplinarian) of Latin American area studies also had the unintended consequence of opening a space to challenge disciplinarity itself. For the historical social sciences, the novelty was to combine nomothetic and idiographic conventions by studying the history of "peoples without history" (Wolf 1982). The attempt to decipher the laws of motion of allegedly traditional societies presented anomalies and challenges. The solution that was offered to the epistemic and political dilemmas involved in this novel task was modernization (developmentalist) theory (Escobar 1995).

In the imagination of Latin American area studies, the vast territory to the south of the Rio Grande (the new frontier) was represented as the past of the United States, which was supposed to be the future of Latin America. A new modality of imperial/colonial discourse translated the colonized subject from the savage or the native into the underdeveloped. But this power-ridden and ideologically based enterprise enabled the creation, organization, and institutionalization of large-scale knowledge-producing apparatuses concerned with Latin American histories, societies, politics, cultures, and languages. It promoted academic institutions and think tanks, both in core countries (the United States, Great Britain, France, and Germany) and in Latin America and the Caribbean (especially in Brazil, Mexico, Argentina, and Chile). This knowledge industry partly facilitated the transformation of the cultures of scholarship and the university system in Latin America, where social science gained more prominence from the 1960s onward with the organization of regional institutions such as the Facultad Latinoamericana de Ciencias Sociales (FLACSO), with its multiple regional centers.

Beginning in the 1970s, a critical strand of Latinamericanism grew in the United States in exchange with Latin American radical intellectual and political cultures. The most influential contribution from Latin America was dependency theory. It was a challenge to modernization theory and its implicit restatement of the concept of the civilizing mission vis-à-vis what was now called the Third World (instead of the Orient or the tropics). *Dependentismo* was a perspective from the 1950s of Latin American social scientists trying to articulate new theoretical terms of discussion and to explain and explore solutions to the historical and structural foundations of the condition of political, economic, and cultural inequality of Latin America in the capitalist world-economy. Simultaneously, and as importantly, dependentismo was a critique of the dominant forms of analysis and political strategies of the Latin American Communist parties, which the *dependistas* argued reflected the same epistemologies as North American modernization theory, which ultimately led to not-too-different political strategies.

Ever since the founding of the U.S.-based Latin American Studies Association in 1964, there was a fracture between the Latinamericanism of Latin American area studies and the critical brand of Latinamericanism that worked in collaboration with Latin American dependency theories. However, the early critiques and challenges of these radical U.S.

Latinamericanists still remained within the general framework of Western structures of knowledge.

In 1991, New York–Puerto Rican political scientist Frank Bonilla reflected on this experience in "Brother Can We Paradigm?" This was an article in which he sought to address from a Latino perspective the initial call of the Gulbenkian Commission for the Reconstruction of the Social Sciences. He recounted organizing a 1973 seminar at Stanford University, "Structures of Dependency," that brought together Latin American dependency theorists with U.S. Latino scholars. At that time already, the project of Latin American area studies was in deep crisis, which, he observes, coincided with the emergence of ethnic and racial studies in the world-historical conjuncture of the "revolution in the world-system" (Arrighi, Hopkins, & Wallerstein 1989; Wallerstein 1991b) of the late 1960s.

Bonilla's seminar was an early attempt to link the group promoting the emerging Latino studies (especially Chicano and Puerto Rican critical social scientists) with Latin American dependency theorists. However, some of the same critiques of elitism and lack of recognition of the knowledge produced by racialized minorities and working-class communities that the movements for Chicano studies and Puerto Rican studies were making of U.S. university-based knowledge were also made of Latin American dependency theorists. It was charged that the dependency theorists were framing the issues only in terms of imperialist domination and the need for national-based critical theory and revolutionary agency without either problematizing the questions of the authority of academic knowledge or recognizing the importance of subaltern knowledge.

At this time, postwar (Cold War) area studies found itself under attack from both the mainstream and antisystemic forces. Some mainstream scholars criticized it for diluting the disciplines because it lacked a clear epistemological object and used a too loose comparative method. It may be that, in the period of so-called détente, there did not seem to be the same urgency for area research. But area studies was also under fire from the movements for decolonization in the periphery as well as movements for democratization in the core, which criticized it for being another version of cultural imperialism and scientific colonialism. Nonetheless, area studies played a significant role in carving a place for multidisciplinary knowledge, developing "non-Western" regions as legitimate objects (but not yet subjects) of inquiry, and promoting regionally based research agendas.

What we call Latino studies today evolved from the intellectual, institutional, and political spaces opened by the Chicano and Puerto Rican movements in the late 1960s and early 1970s. Chicano-Riqueño radicals for the most part had a Third-Worldist transformative vision in which national liberation implied the construction of politicized identities by retrieving memories of resistance. They sought to decolonize U.S. institutions by combating in-built ethnoracial and class inequality and by promoting Latino power and self-representation. Late-1960s Chicano and Puerto Rican revolutionary nationalisms emerged in the wake of the explosion worldwide of anticolonial movements and theories in the Third World[2] and were informed by the understanding of Chicanos and Amer-Ricans as "conquered minorities" and colonial subjects. This colonial model was the dominant paradigm of the first programs of Chicano and Puerto Rican studies.

In general, ethnic and racial studies, as well as women's studies (and later gay and lesbian studies), were the product of synergized struggles (of students and academics, community organizers, and activists of all sorts) to open and democratize the universities. The demands of access for the excluded (working-class, racialized minorities), the democratization of university governance, the collectivization of the production of knowledge, the creation of a participatory pedagogical process, and the transformation of curricula to include the histories and cultures of oppressed peoples gave shape, in the context of higher education, to the politics of decentralization, collective leadership, and local power that inspired the new social movements with a radical democratic ethos.

This early moment of what we now call Latino studies was also characterized by a search for an organic relationship between the university and subaltern communities, an active relationship between theory and practice, and an assertive campaign for the decolonization of knowledge. The critique of knowledge was threefold: first, of the negative and stereotypical representations of Latinos in the U.S. academy and mass culture; second, of the capitalist and colonialist character of U.S. university-based knowledge; and, third, of the marginalization and erasure of the histories and experiences of colonized peoples and subaltern minorities from the dominant modes of representation.

These movements and their political and intellectual agendas led to creating programs of research and instruction, recruiting students and faculty from sectors that had barely been present before in academic institutions, effecting some change in the curricula and personnel of the disciplines, and opening the debate on standpoint epistemologies and the politics of recognition and representation that were to characterize the culture wars of the 1980s and 1990s. They also contributed to the framing of U.S. history and social realities beyond the optic of the nation-state as the primary unit of analysis, by developing a transnational approach to the histories and politics of their communities. In spite of these achievements, the early project and the practice of ethnic studies also had serious shortcomings: its definition of peoplehood highlighted race and class and marginalized gender and sexuality, and its critique of the structures of knowledge did not articulate a full-fledged epistemological and methodological alternative to disciplinary knowledge that could transcend the divide between the sciences and the humanities and that could coin new discursive categories and terms of intellectual discourse. To some extent early Latino studies can be characterized as new political wine in old epistemological bottles.

Once the antisystemic movements entered into a downswing in the 1970s after their moments of triumph in the late 1960s, the maximalist hopes for the relationship between theory and practice and for collaboration between the university and the communities were frustrated. The political, ideological, and intellectual contradictions within Latino studies were exacerbated. And a trend toward marginal professionalization and institutionalization gained strength. The revolutionary aura of ethnic studies had disappeared along with the theory of internal colonialism, giving place to a new contested terrain within Latino studies, in which one of the main debates was to be between those who emphasized the critique of power and knowledge from feminist, queer, and cultural studies perspectives on one side and both Marxist social science and policy-oriented empiricist outlooks on the other.

What followed within U.S. universities seemed like a revolving door in which Chicano and Puerto Rican studies struggled to survive budgetary austerity and endemic academic hostility (especially in the public universities), at the same time that Latino studies came to be in vogue and different sorts of fusions between Latino and Latin American studies were attempted (especially in the private research universities). The rise of Latino studies had two distinct sources that unwittingly coincided in their effects. At one end, the vogue for Latino studies was the result of cultural struggles (educational, art, media) for diversification and democratization of public culture, and in particular of university-based knowledge, reinforced by a new wave of student politics inspired by identity politics. At the other end, Latino studies resulted from the increasing commodification of race and ethnicity as cultural capital for liberal knowledge industries competing in the academic market. That is to say, a deeper corporatization and globalization of U.S. knowledge-oriented institutions (universities, foundations, think tanks) was also an important force facilitating the fusion of Latino ethnic studies and Latin American area studies.

As we have seen, Latin American area studies and Latino ethnic studies have distinct genealogies (Caban 1998), given their different timing in the histories of the modern world-system (rise/decline of U.S. hegemony), enabling agencies (powerful institutions/antisystemic movements), their main political goals (modernization and crisis management/democratization of the university system), and epistemological projects (interdisciplinary research with a comparative method/decolonization of knowledge production and pedagogy). These alternatives were not only between but also within each of the organizational structures, since both of them were from the beginning contested terrains.

There existed a critical variety of Latin American studies, both in the United States and Latin America. On the other hand, following the decline in the role of antisystemic movements and the consequent institutionalization and professionalization of Latino studies, the field was partly transformed into a marginal academic sector struggling to survive neoconservative attacks and neoliberal budgetary retrenchments.

Latino studies developed differently in the various regions of the United States—Southwest, Northeast, Midwest, and Southeast. One factor was the specific history of each region as an imperial contact zone:[3] the 1846–1848 Mexican-American War in the Southwest; the 1898 war in the Northeast; the 1959 Cuban Revolution in the Southeast. Other factors included the distinctive timing and composition of immigration, the differential modes of incorporation and community making (class formation, social movements, leadership), and specific academic histories (intellectual, institutional, political).

These regional differences are more relevant for understanding the main actors, key populations, and principal themes of research of Latino studies, but they also affected the configuration of Latin American studies. For instance, it is no accident that the most important center for Caribbean studies in the United States is at the University of Florida at Gainesville, while the main center for the study of U.S.-Mexico relations is at the University of California at San Diego, and the only independent PhD program in Latin American studies, which concentrates on Mexico, is at the University of Texas at Austin. But after the relative decline of area studies and the rising claims of the excluded to carve out spaces in the university system in the 1960s and 1970s, it was the Latino studies movement that created these regional academic niches. In fact, as we have seen, until the 1980s and 1990s most of what now is called Latino studies was known as Chicano studies in the Southwest and Puerto Rican (or Black and Puerto Rican) studies in the Northeast. But in the late 1980s and 1990s, in light of the growth, diversification, and dispersion of migrations from Latin America and the Caribbean to the United States, we witnessed the emergence of smaller scale movements advocating *Latino* studies as well as the organization of nationally defined programs such as the Institute for Dominican Studies at the City University of New York and the Center for Cuban Studies at Florida International University.

The most significant thing about Latino studies is that the field emerged as an explicit effort from below (working-class racialized communities) to open up and democratize/decolonize university-based relations of knowledge production, governance, and pedagogy. As we have seen, this agenda did not achieve its full potential because of the political limitations created by the exhaustion of the antisystemic movements in the 1970s and 1980s, as well as because of the epistemological shortcomings of a critique of knowledge in early Latino studies that did not advance beyond addressing the question of exclusion (by race and class), denouncing negative representations of colonial subject peoples, and criticizing dominant science as ideological, utilitarian, and disempowering to subaltern subjects. This incipient critique of the dominant structures of knowledge was important but lacked a more fundamental analysis and fleshed-out alternative. Later, however, the seed that was planted by the insurrection of excluded knowledges (African-American, Asian-American, Latino, Native American, women, gay, and lesbian)[4] in the 1960s and 1970s was replanted and cultivated by the movement for cultural studies in the 1980s.

Latina/o studies and its scholars (as well as other ethnic studies programs) played an important role in the movements and in the institutionalization of cultural studies (Grossberg, Nelson, & Treichler 1992; S. Hall 1992). They prepared the ground for cultural studies by articulating a notion of culture as constitutive of (and constituted by) power, and by claiming a space for transforming the structures and institutions of academic knowledge from the standpoint of the excluded and marginalized in social structures of power. But the links between knowledge and power were elaborated in a more systematic and fundamental critical way by cultural studies along several lines: the question of technoculture (Aronowitz 1993; Penley & Ross 1991), the arguments for standpoint epistemologies (Alcoff & Potter 1993), and the call for a trans(post)disciplinary reorganization of theoretical practice and research activity, this also entailing a transgression of conventional boundaries between the two cultures (sciences and humanities). The relationship between Latino and cultural studies has been a rather complex one in which reciprocal influences have combined with mutual challenges.

Chicana/o intellectuals and activists coined notions such as "border" and the "politics of location" that became central in the rhetoric of cultural studies. Both notions were key to a new strand of intellectual labor that not only accentuated ethnonational/racial and class oppression and redress but also engaged in a deconstruction of *chicanidad/ latinidad* from the perspective of those—women, gays, and lesbians— who were excluded from and marginalized in the dominant narratives of Chicano/Latino peoplehood (Dominguez 1989; Wallerstein 1991a). The concept of internal colonialism (Blauner 1972; Barrera 1979; Stavenhagen 1973), which had served as a crucial element in the arguments of the early Chicano revolutionary nationalists (Acuna 1972), was replaced by the notion of borderland (Anzaldúa 1987),[5] more appealing at a moment in which intellectual and political concern focused on the play of differences, translocal connections, and hybrid states of being and definitions of self. But for Chicana/os, borders and borderlands were not simply post-structuralist discursive spaces and theoretical metaphors to deconstruct master narratives and signify the multiple mediations of identity/difference. They were first of all a material geographic imposition

located on the two shores of the Rio Grande, with concrete politicoeco-
nomic and existential effects on the conditions of life and death of people of Mexican descent. The tensions between Chicana/o and cultural
studies are more clearly revealed in the concept of the politics of location
developed by Chicana feminists to conceptualize the multiplicity of subject positions (class, gender, sexuality, race, ethnicity, generation) embodied by individual subjects as a complex (and somehow contradictory)
grid of domination and subordination, and to theorize this locus of enunciation as an epistemological location. Hence, here the critique of the
Eurocentric, particularistic character of Western universalism is extended to all forms of knowledge, including those produced within Latino,
women's, gay/lesbian, and cultural studies. The question remained, as
we shall see, whether there exists (or whether we should look for) criteria of universalism and objectivism, or whether there can only be localized and situated partial knowledges.

Puerto Rican studies made an important contribution to the emerging constellations of knowledge. By focusing attention on the enduring
and translocal colonial nature of U.S.–Puerto Rico relations, it pushed
the project of Latino studies (and by implication those of American studies, cultural studies, and Latin American studies) beyond the nation-
state and within global, regional, and imperial frameworks. In spite of
the theoretical and political limitations of the 1970s' theory of internal
colonialism, the persistence of the colonial status of Puerto Rico and the
everyday lived experience of colonial situations by U.S. Puerto Ricans
made the question of coloniality an imperative for Puerto Rican studies.
Thus, it was Puerto Rican studies that undoubtedly raised most clearly
the problem of the importance of empire in U.S. academic discourse.
Nonetheless, ironically, Puerto Rico and Puerto Ricans had been notice-
ably absent from the current boom of studies on colonial discourse and
postcolonial theory. This is symptomatic, on the one hand, of the lack of
discussion of the U.S. empire and its colonial subjects in mainstream
postcolonial theory, but it is, on the other hand, also a product of a
relative stagnation of the theoretical and political treatments of the question of coloniality in Puerto Rican studies (Santiago 1994).

Postcolonial studies and cultural studies have presented challenges
(both from outside and from within) to Latino studies, pushing the field
toward a more fundamental analysis of its epistemological underpinnings. Thus, since the mid-1980s, the newly defined field of Latino studies emerged as a contested terrain between advocates of ethnic studies
(mostly old-guard social scientists, including Marxists) defending empirical social-historical research and a return to community populism
and proponents of Latina/o cultural studies (many of them newcomers
from the humanities) practicing a combination of textual readings, sociohistorical analysis, and ethnographic-based local knowledge. The latter,
in exchange with currents in postmodern and postcolonial theories, de-

veloped an epistemological critique of Western structures of knowledge predicated on a concept of the knowing subject as sovereign, abstract, disembodied, all-rational, self-righteously moral, and omnipotent. The argument is that this subject of Western science and instrumental rationality is by definition an ideal representation of a White, male, heterosexual, capitalist, imperial gaze, and as such an epistemic premise for the exclusion of women, homosexuals, "lesser races," natives, and subaltern classes from the production of legitimate knowledge. Therefore, they argue, there is a need to criticize the modes of legitimation and mechanisms of reproduction of the authority of Western knowledge, to replace them with a decolonized epistemology based on a conception of subjects as embodied, gendered and eroticized, colonized/racialized, and therefore inscribed in power relations. This more fundamental critical stance on the epistemic principles of Western knowledge involved a challenge to the hegemony (and to the alleged value-free objectivity) of science, as well as ending the division between the sciences and the humanities. This kind of critical discourse has become an explicit agenda for the new voices in Latina/o cultural studies (Alarcon 1991; Aparicio 1997; Sandoval 2000; Saldivar 1991; Saldivar-Hull 2000).

In these new intellectual scenarios of Latino studies, nation and ethnicity, though still important, are no longer the only (and in some cases not even the main) locations from which to enunciate a position of knowledge. The question of identity has been deconstructed and complexified in a way that no single marker of identification is necessarily dominant, at the same time that peoplehood is no longer identical to nationhood. To a large extent, in most of the analysis informing curriculum and instruction in Latino studies, national identities have continued to take priority over Latino panethnicity, but the field has also seen opposition to nationality as the main marker of identification and developed more the implications of its always transnational outlook for transcending the nation-state as the main unit of analysis and principal terrain for political and cultural struggles, emphasizing instead the local, the world-regional, and the global. The relative decline of nationalism in Latino analysis allowed "race" and racism to emerge as primary categories of analysis in Latino studies, including the specificity of the Afro-Latino experience. The growth of critical racial studies and critical legal studies as areas for Latino scholarship opened yet another intellectual space for the interpretation of Latino identities as racialized modes of affiliation and subjectivity. However, even though Latino studies always involved interpretations of *latinidad* (or *chicanidad, puertorriqueñidad,* etc.) as racialized categories, for the most it did not analyze latinidad as a racial formation (or racial formations within latinidad). Consequently, Latino studies has not in general articulated a serious critique of the racial substratum of Western thought and the racial logics of Western philosophy, as has been done in African and African-American studies, and in postcolonial theory.

Latin American studies has also become an arena of intellectual and political debate. First, alongside the old guard of Latinamericanists using modernization theory or pursuing language studies and the old cadre of Marxist social scientists, new voices and paradigms in Latin American area studies are being heard. Some anthropologists and historians are developing a critical perspective on hemispheric, national, and local relations of power in the Americas. Their interpretation of imperialism is not only geopolitical and economic but also involves transculturation and unequal cultural exchange between North and South (Pratt 1992b; Joseph et al. 1998). This tendency to take the whole of the Americas (instead of the two subhemispheric regions and/or the nation-states) as the immediate unit of geohistorical analysis is also gaining momentum within the field of American studies (Kaplan & Pease 1993; Belknap & Fernandez 1998; Saldivar 1997).

Second, a Latin American subaltern studies group has been organized by U.S.-based academics (both North Americans and Latin Americans), following the model of the South Asian subaltern studies group. This group, composed mostly of literary scholars and radical historians, articulated more a political than an epistemological challenge. Their project for the empowerment of Latin American subalterns (ethnoracial, gender/sexual, class) was in opposition to the intellectual project of a Latin American cultural studies that had been maturing in semiperipheral metropolitan centers (like Mexico City, Buenos Aires, Santiago de Chile, and Montevideo). This latter perspective presented itself as a critique of both modern and postmodern structures of knowledge in core capitalist centers from the standpoint of what some of the critics call peripheral postmodernism (Achugar 1994; Garcia-Canclini 1990; Sarlo 1994). The distinctive mark of this sector is their "postmodern" analysis of postdevelopmentalist neoliberal Latin America, arguing for a breakdown of the master narratives of nationalism and world-regional revolution. Although engaged in an exchange with European and U.S. postmodern theories, these scholars also took critical distance from them from a Latin American perspective (Richard 1989, 1993, 1994).

Finally, yet another group of Latinamericanists, most of them Latinos and Latin Americans, located throughout the Americas, have argued in favor of a Latin/o American subalternism from the epistemological and political standpoint of a critique of coloniality within a world-systemic perspective.[6] Three analysts—U.S.-based Argentinian critic Walter Mignolo, Peruvian social theorist Aníbal Quijano, and Mexico-based Argentinean philosopher Enrique Dussel—share a world-systems perspective of modern history as a capitalist world-economy that emerged in the long sixteenth century, a reinterpretation of modernity from the standpoint of its colonial undersides, and a critique of Eurocentric (or Occidentalist) regimes of knowledge.

Quijano coined the notion of the coloniality of power as a key concept to explain the entanglement of four world-historical processes: first, the rise of historical capitalism based on capital's exploitation of a diversity of forms of labor; second, the reorganization of political bodies primarily as nation-states engaged in unequal relations of imperial/colonial domination; third, the emergence of modern racial and ethnic hierarchies of classification and stratification corresponding to the remaking of gender hierarchies; and, fourth, the creation and institutionalization of Eurocentric rationalities and structures of knowledge. Quijano theorizes a relationship between world-capitalist modern/colonial structures of power and structures of knowledge, and between the subordination of subjects and the subalternization of knowledges. This correspondence between the coloniality of power and the coloniality of knowledge is manifest not only in the geocultural and institutional allocation of academic power and the authority of knowledge but also in the very logic and method that configure Western epistemologies. For Quijano, Eurocentric rationality is instrumental, dualist, and lacking a notion of intersubjectivity. It is based on an organic functionalist concept of the totality. Quijano calls for the decolonization of knowledge by dismantling dualism and instrumental reason and refusing the geocultural authority of Eurocentric rationality.

Dussel, a philosopher committed to a project of liberation, is more concerned with a critical engagement with the Western philosophical tradition from a Latin American peripheral location in the world distribution of legitimate knowledge. Dussel sees "the world-system as a philosophical problem," in the sense of being the ultimate ontological framework for the emergence of structures of knowledge allocating differential epistemic power, authority, and legitimacy to central and peripheral locations. He seeks to articulate a truly global interpretation of the modern, grounded in what he calls the reason of the other, in the histories of those others (workers, women, colonized and racialized subjects) who had been excluded from the false universalism of Occidentalist rationality. For Dussel, what he calls transmodernity opens an intellectual space to conceptualize at once the diversity of modern logics and a political and epistemological critique of all the forms of domination and exploitation that are constitutive of the modern world-system.

Dussel, like Quijano, theorizes the Spanish conquest, colonization, and enslavement of Amerindians and Africans as an ontological foundation of the episteme of what he calls the first modernity. Hence, the rise of the Atlantic system and the creation of Ibero-America as a world-historical region is conceptuallized as constitutive of modern regimes of power and knowledge. This sets a foundation for a critique of the colonial undersides of capitalist modernity and its Occidentalist imaginary.

Mignolo builds from both Dussel and Quijano, as well as from Wallerstein, to develop his analysis of the problematics of knowledge in the colonial horizon of modernity. In *Local Histories/Global Designs: Coloniality, Subaltern Knowledges, and Border Thinking* (2000), he explicitly engages the analysis and proposals made by the Gulbenkian Commission in *Open the Social Sciences* (1996). He seeks a dialogue between world-systems analysis and postcolonial theories, hoping to achieve a synthesis. Mignolo makes four general arguments that are directly pertinent to our discussion of the structures of knowledge. The first, following Quijano, is that coloniality is not accidental or marginal to modernity but a central aspect of modern regimes of power/knowledge; consequently he renames the system as modern/colonial. The second is that, consequently, Occidentalism (as the dominant imaginary of the system) is characterized by a necessary subalternization of non-Western modes of knowledge, and thus the systemic silencing and erasure of other knowledges is a fundamental feature of the Western structures of power/knowledge. The third is that there is a historical sequence in the occidentalization and subalternization of other knowledges developing from sixteenth-century Occidentalism and eighteenth- and nineteenth-century Orientalism to post-1945 area studies. Here, the creation of area studies is seen as a keystone in the late modern geocultural global division of knowledge between three worlds (Pletsch 1981) where the first world (the capitalist core) allegedly produced science, the second world (the Soviet bloc) produced ideology, and the third world (the rest) produced culture. The fourth is his proposal for a border gnoseology or border epistemology. Mignolo defines border gnoseology as a perspective from the external (i.e., colonial) borders of the system (against the internal margins) and defines its rationality as countermodern (or what he calls postcolonial and/or subaltern) reason.[7]

There have come to be two main issues in the epistemological and political challenges to Latin/o American studies. The first is race. The movement for Latino studies has been part of a challenge to the Western structures of knowledge from the standpoint of those racially and ethnically excluded from the dominant standards of rationality, truth, goodness, and beauty. This is found in the antidiscrimination language, and the self-affirmation of Latinos as racial and/or ethnic subordinate groups, that is present in most versions of Latino discourse. However, for the most part, there has not been in Latino studies a critique of Western epistemology from the standpoint of race and ethnicity. For instance, there have been demands of Latino/a scholars for the modification of mainstream canons (in art, history, and literature) to include Latina/o cultural production and to promote Latina/o self-representations. But these claims are generally made in the name of peoplehood, usually conflating notions of race, ethnicity, and nationality, with the analytical consequence that race as a primary category of modern/colonial identification and racism as a world-historical system of domination are diluted and marginalized.

It is only with the entry of critical race studies and critical legal studies into Latino studies (Delgado 1995) on the one hand, and of the analysis of the coloniality of power in Latin/o American studies on the other (Quijano 1991), that race became a primary category, and therefore a foundation for a critique of modern regimes of power and knowledge. In a similar but distinct way, African/Afro-diasporic and postcolonial thinkers have recently developed critical theories of Western thought and philosophy (including epistemology) in various ways, from exposing the racist contents and subtexts in the Enlightenment philosophies of history and notions of the self and demonstrating the colonial/racial undersides embedded in the positivist/holistic imperial epistemological gaze (Said 1978; Spivak 1993), to showing the substantive racial elements in the theory and practice of Western political philosophies and epistemologies (Mills 1997; Goldberg 1993).

As we have seen, in the case of Dussel, Quijano, and Mignolo, the critique of knowledge also entails deconstructing the very logics, categories, and rationalities of Occidentalism and reconstructing the structures of knowledge with other logics, concepts, and rationalities. In this project the concept of race is fundamental (Quijano 2000b; Santiago 1994), because as the main specific criterion for colonial difference (Chatterjee 1993), race is a key marker of selfhood and inequality (domination and exploitation) and therefore constitutive of (and constituted by) the structures of power and the structures of knowledge.

Colombian philosopher Santiago Castro-Gomez argues that we need yet another step for a fundamental critique of the Western episteme. For Castro-Gomez this third critical move is best represented in the efforts by postcolonial theory to interrogate and challenge the imperial/colonial character of the very epistemological principles and procedures of Western structures of knowledge, the practical implication being the need for a redefinition and reconfiguration of world epistemologies not only through an immanent critique (from within) of Occidentalism but also via an external critique. It is imperative to move beyond the self-referential terms of conversation given by Occidentalist self-reflections on the histories and problematics of knowledge.

The representatives of the postcolonial moment in world-systems analysis—Dussel, Quijano, and Mignolo—are generally informed by a notion of the sociohistorical totality as "structurally heterogeneous" (Quijano 2000a), heterarchical (Grosfoguel 1995), and always constituted in exchange with constitutive externalities (Dussel 1995, 1996, 1998; Mignolo 1995, 2000). They imply an understanding of the structures of knowledge as a multilayered process of entanglement operating simultaneously, but with significant degrees of relative autonomy, at the local, national, regional, and global levels—hence, both the analysis of world-historical regional structures of knowledge and the challenge to institutionalized knowledge by the unacknowledged and devalued knowledges of subaltern classes. This constitutes a crucial contribution of the new critical Latinoamericanism to contemporary challenges of dominant regimes of power and knowledge, as well as to world-systems analysis in particular.

This search for a plural, relational, and contextual universalism as a product of a dialogue between different locations of power and perspectives of knowledge is also part of the project of the Gulbenkian Commission report. But the idea in the Gulbenkian report that the social sciences might mediate the crisis in the structures of knowledge, bringing together the internal challenges to Newtonian scientism from the sciences of complexity and the challenge by cultural studies to the concept of canons abstracted from social context, had not been really considered in the new critical Latinoamericanism. This is partly because in Latin America the social sciences have had a less disciplinary character than in the United States. In addition, critical theories (e.g., dependency theories, social movements) have been central in the overall definition of Latin American social sciences, and a substantial number of social scientists combine intellectual labor with activism, organizing, and political debate. As a result, cultural studies in Latin America, in contrast to the United States, has emerged more from within the social sciences, and its struggle has been, together with a new generation of cultural critics, partly against the conservatism of a Latin American Occidentalist tradition in the humanities. Likewise, the challenges emerging from the current crisis in the natural sciences are primarily developing from within Latin American social sciences (Piscitelli 1992, 1992–1993; Salvatore 1998), and also from intellectuals working with grassroots social movements for environmental justice, alternative development, and economic and cultural rights (Escobar 1995).

The case of Latino studies is different, to a large extent because it is a field that was born inside the United States. As we have seen, most of the early scholarship in the field came from social scientists. In contrast, a considerable percentage of the criticism from Latino cultural studies of Western regimes of power knowledge, but also of the shortcomings of the heterosexism and nationalism prevailing in the first versions of Latino studies, came from scholars in the humanities. As already discussed, perhaps the most significant contribution from Latino studies to the critical discourse on the Western structures of knowledge is the notion of the politics of location and its implications for a complex and nuanced articulation of the feminist principle of standpoint epistemologies searching to deconstruct the White, masculinist, imperial gaze of the allegedly neutral, rational, and scientific ideal Western subject of knowledge, from the perspective of embodied, sexed, classed, raced, and ethnicized subjects concretely located in time-space. This in itself is not necessarily an argument against ontological realism or epistemological objectivism, but rather, in its most philosophically conscious versions (Alcoff & Potter 1993), is a defense of a new self-reflective objectivism, one that acknowledges mediations of power, interests, and desire, in the production of meaning under particular regimes of truth. They go beyond simple-minded searches for authenticity or for an unmediated authority of experience. Latina/o arguments—mostly feminist, gay, and lesbian (Anzaldúa 1987; Sandoval 1991; Muñoz 1989; Perez 1999)—that relate the politics of knowledge to the politics of location analyze how the multiplicity of subject positions and the myriad of lived experiences of particular agents (gendered, sexed, classed, raced) existentially guide their cognitive and affective dispositions for knowledge. Mignolo builds from this Latina/o studies discussion of the relationship between structures of feeling and structures of knowledge to develop the notion of the epistemological loci of enunciation in order to give a more general epistemic meaning to the notion of location. This illustrates how the ambitions and perspectives of U.S. Latino critiques of knowledge have been more local/particular than the discourses of Latin American postcolonial theory that sought to articulate macronarratives from the colonial horizons of modernity (Mignolo 2000).

Where postcolonial approaches from Latina/o and Latin/o American studies converge most is in questioning and challenging Western regimes of power and knowledge from the perspectives of subalternism. However, this is a thorny question for two fundamental reasons. One is the gap that exists between the institutions and means of representation of those who speak on behalf of the subaltern and the modes of production and communication of knowledge of the subaltern classes themselves. The second is the difficulty in defining and specifying precisely who are the subjects of what kind of subalternity, given the complexities and contradictions of modern structures of power and knowledge. These questions have given rise to interesting debates about the differences and difficulties of dialogue between subaltern knowledge and academic subalternism (e.g., in Latino ethnography as insider/outsider) (Chabram 1990), and on the relative and relational nature of subalternity as a category and subaltern studies as a perspective (Coronil 1996, 1997; de la Campa 1999; Moreiras 2001). In any account, the lower echelons of subalternity (the marginal others or the radical alterities) are inhabited (at the global, regional, national, and local levels) by the racialized (as well as gendered and eroticized) stratum of labor (Santiago 1994), which in the Americas is mostly composed of African-American and Amerindian peoples. This does not mean an attribution of a moral and/or epistemic superiority to this level of subalternity but is a way of showing how the synergy of locations of subordination unequally allocates power, knowledge, and wealth.

This is also a way of revisiting the question of race and knowledge in Latinamericanism, because an important implication is that in Latin/o American studies there is a double articulation between race and subalternity. First, there is a very general division between the so-called Anglo-Saxon race and the so-called Latino (or Hispanic) race; and, second, there is a division between Euro-Latinos on one side and Afro-Latinos, Amerindians, Asian-Latinos, and the like on the other side. Thus, we have at least two distinct layers of subaltern modernities (Coronil 1996) and subaltern knowledges, the implication being that the relationship between race and subalternity in the Latinamericanist critique of Western structures of power and knowledge should involve a double critique (Mignolo 2000) of racialization/subalternization—both of the core Occidentalist dominant regimes of power/knowledge, and of peripheral Occidentalism and its institutionalization and disciplinarization of knowledge to the exclusion of its own others. An important political extrapolation from this subaltern perspective is the question of agency for change or, in other words, the need for a new wave of social movements from below to challenge and transform the structures of knowledge not only from within its institutions but also from outside its formal domain. In this moment of world-systemic crisis there is a bigger potential, but for the very same reason a greater challenge, for the emerging movements to become antisystemic. Latin/o American postcolonial theories imply a political project in that the decolonization of knowledge is conceived as a crucial ingredient in the democratization of power relations on a world scale.

Notes

1. The concept of transculturation was coined by Fernando Ortiz ([1940] 1963) to conceptualize the unequal exchange of culture and capital between different classes and ethnoracial groupings in the making of the Cuban nation. Two important extrapolations and elaborations of the concept for a Latinamericanist postmodern and postcolonial critique are Coronil (1997) and Pratt (1992b).

2. There are different visions of the notion of the "Third World." Escobar (1995) sees it as a product of post-1945 developmentalist discourse. Stam and Shohat (1994) see it as a way of naming the struggles for self-determination and decolonization in the periphery of the modern world-system.

3. For the notion of imperial contact zone as a field of uneven development and unequal exchanges of all sorts (economic, political, cultural), see Pratt (1992b).

4. This formulation is a combination of Foucault's insurrection of subjugated knowledges (1977) and Wallerstein's notion of excluded knowledges (Gulbenkian Commission 1996).

5. Indeed, the use of the concept "border" as a political signifier of the condition of Chicanidad can be traced at least to the early 1970s. It is exemplified by the slogan of the radical organization CASA, "We are a people without borders." See Garcia's biography of Bert Corona (1989).

6. We have been using the terms "global" and "world" interchangeably thus far, but they have different meanings and signify distinct theoretical traditions. The language of globalization is usually used in relationship to the tremendous increase in circulation of capital, peoples, and representations in the last twenty-five years, generally without recognizing that globalization itself is a process that is coterminous with the modern world-system since its emergence in the long sixteenth century. A

world-economy and a world-system is a geohistorical unit of social life. There have been a number of these in human history, but the only one with a global reach is the modern world-system (Arrighi 1994; Braudel 1984; Wallerstein 1974).

7. Mignolo's argument (2000) is partly framed in the terms of world-systems theory insofar as he assumes the modern world-system as a historical and social unit of analysis. His critique of Western structures of knowledge also has much in common with the Gulbenkian report's search to overcome the distinction between the sciences and the humanities, transgress disciplinary boundaries, and supersede the legacy of Newtonian objectivism, Cartesian dualism, and Western progressivism. They also coincide in the call for a contextually bounded, plural universalism based on broad-based dialogues crossing all sorts of national, class, racial, gender, and sexual borders. A crucial difference is Mignolo's contention that the modern world-system looks different from the perspective of coloniality and the implications this has to theorize the epistemological locations of the subjects of knowledge, and to conceptualize alternate structures of knowledge in geocultural locations apart from Europe and the United States. This will entail a decentering and remapping of the ways in which we see the cultures of scholarship, subaltern knowledges, and their possible cross-fertilizations. This is part of what might be called the postcolonial moment in world-systems analysis.

12

Environment and Ecology: Concepts and Movements

Sunaryo

The relationship between human beings and the rest of the biosphere, living and nonliving, has become a major worldwide concern academically, socially, and politically since 1945 (Lee 1996; Gulbenkian Commission 1996; Demeritt 1994; Eckersley 1992; McCormick 1995), although various ecological and environmental movements have existed since at least the late nineteenth century. Indeed, the "environment" was long thought to be a, if not the, primary causal influence effecting human history (environmental determinism). From the late nineteenth century onward, concern for the environment continued, but in terms of "conservation." However, a more holistic conception of the world, ecology, which stressed not just physical factors but entire systems of relations in which species and nonliving surroundings interacted to form a single unit, was also emerging, albeit in the shadow of the conservation movement. The two could even be linked, and "by the time of the 1930s and 1940s, ecology was being hailed as a much needed guide to a future motivated by an ethic of conservation" (Woster 1991: 1). Such a perspective had become the foundation of political ecology, which, however, remained a preserve of a small section of the European and American intelligentsia until after the First World War (Bramwell 1994: 2).

In the last quarter of the twentieth century, these movements took a different turn, becoming much more fundamental in their critiques and more aggressive in their social action. We shall nonetheless distinguish between environmental and ecological movements. We shall call those movements environmental that concentrate on conserving the environment. They often act via philanthropy (e.g., World Wildlife Fund) or otherwise generally employ traditional political means of achieving their ends (e.g., Sierra Club, National Audubon Society). We shall call those movements ecological that attempt to act on holistic views of the biosphere by raising questions about the relation of economic, biological, and moral priorities. These movements often engage in direct action (Greenpeace, Earth First!) (Dobson 1995; Merchant 1992; Bookchin 1991). The ecology movement is further divided by a segment that insists on what they call "deep ecology" (Devall & Sessions 1985; Drengson & Inoue 1995), a position that challenges any primacy of human beings over other life forms in the biosphere.[1]

The early movements of the late nineteenth century were largely conservative movements in the double sense that their primary aim was conservation and their politics tended to be conservative. Their major concern was to preserve wildlife (Fox 1985) rather than the ecosystem as a whole. As late as 1945, it seemed inconceivable that environmentalism would transform itself into a major political force (Wall 1994: 35), and one that tended to be on the left rather than politically neutral.

The traditional environmentalist movements were split between two emphases, landscape preservation and scientific conservation. The leading protagonists of these two positions in the United States were John Muir, the first president of the Sierra Club, elected in 1892 (H. Smith 1965); and Gifford Pinchot, whose appointment as the first chief of the U.S. Forest Service by President Theodore Roosevelt marked the political victory of the conservationist position (Miller & Sample 1998).

The conservationist position came under attack after 1945 by Aldo Leopold and others who argued that its successes were at the expense of the overuse of land, the pollution of the air and water, and the loss of wildlife and that this was therefore a one-sided victory (Shepard 1969: 8–9). Leopold asserted that the fundamental weakness of the approach lay in making economic considerations the only motive for conservation. He pointed out that many types of terrain—marshes, bogs, dunes, and deserts—do not have obvious economic value but do have ecological value (A. Leopold [1949] 1969: 407). Adopting economic self-interest as the sole basis of conservation could only lead to the eventual disappearance of many species that have no immediate commercial value, even though they are important elements of the biosphere. The conservation approach takes for granted that the commercial biotic circle can function without the noncommercial one. "The fallacy the economic determinists have tied around our collective neck, and which we now need to cast off, is the belief that the economic determines all land-use. This is simply not true" (A. Leopold [1949] 1969: 415).

Leopold interpreted the history of ethical thinking as evolutionary, stressing how the conception of ethical agents has expanded individual-ism to encompass societies and social goods. He argued that current needs and values demanded further extension of the "moral communi-ty" to include the biotic community—the ecosystem—and its members (Cuomo 1998: 42). Leopold's central theme was that a healthy, produc-tive, and attractive environment is a precondition of the well-being of humankind (Shepard 1969). This is obvious "when human population and its problems are subjected to the same ecological scrutiny as now is applied to the study of other animal populations. With all his technolog-ical miracles, man is still basically an animal, with all the natural needs, reactions, and dependencies of an animal" (A. S. Leopold 1969: v).

Although Leopold challenged the basic premise of land exploitation, he did not really question the role of the development paradigm and science in the deterioration of nature. His conceptualization of ethical agents, which included societies, ignored class differentiation among its members. In contrast, William Kapp, in *The Social Cost of Private Enter-prise* ([1950] 1970), argued that microeconomic considerations led enter-prises to externalize environmental costs to a third party, the society as a whole, both the current and future generations. Kapp argued that

> the institutionalized system of decision-making in a system of business enterprise has a built-in tendency to disregard those negative effects on the environment that are "external" to the decision-making unit. . . . A system of decision-making operating in accordance with the princi-ple of investment for profit cannot be expected to proceed in any way other than by trying to reduce its costs whenever possible and by ignoring those losses that can be shifted to third persons or to society at large. ([1950] 1970: xiii)

Kapp ([1950] 1970) suggested that the microeconomic principle of individual firm investment, which emphasizes profit, has always been the driving force of environmental destruction. At a macroeconomic lev-el, national accounts, which measure the national economic performance based on economic growth, are misleading, he said, because they ignore the social cost that is left to third parties and future generations to pay.

Many people date the birth of the ecological movement to 1962. This is the year of the publication of Rachel Carson's book, *Silent Spring*. This book provocatively showed the side effects of the application of science in industrial production and also determined the tone for much subse-quent Green political activity with its potent mixture of restrained emo-tion and thorough scientific research.[2] Since then "ecology has been a subversive science in its criticism of the consequences of uncontrolled growth associated with capitalism, technology, and progress—concepts that over the last two hundred years have been treated with reverence in Western culture" (Dobson 1991: 261). The term "ecology" moved from being a term used primarily by scientists[3] to the popular and political lexicon in the 1960s, when a series of books convinced the public that industrial production threatened the well-being of the earth (Russell III 1997: 31). The word ecology seemed to fit the new environmental awareness.[4] Ecology grew from being a minor branch of biology to an interdisciplinary study linking the natural and the social sciences. The transformation of ecological science into a political ecology movement went hand in hand with the rise of the civil rights and anti–Vietnam War movements.

The ecological movements of the twentieth century were immediate responses to the continual environmental crisis (Devall & Session 1985). These movements have solved some of the problems, reformed some of the regulations and agencies that manage the natural resources, and altered popular behavior of some people toward the environment. But for many in the movements, such reforms were inadequate, and they called for a new ecological philosophy (see Spretnak & Capra 1986; Dobson 1995; Merchant 1992). The Green parties, which have been constituted in a number of countries, claim that repair of past errors is insufficient; it is necessary to analyze the causes of ecological degradation and attack the problem at its roots. Greens see industrialized farming as just another example of a blighted way of life that creates a long series of environmental "single issues," such as acid rain, global warming, and holes in the ozone layer. They say that none of these issues can be handled in isolation from the political and economic system that has created them. Greens argue that, since our planet's resources are finite, the only viable society is a "sustainable" society. Greens suggest that none of our societies at the present have sustainability beyond the next few decades. A sustainable society would find ways to minimize the goods, rather than to maximize them as we do now (Dobson 1991: 5).

The ecological movement has faced a serious challenge of perception. As Ernst Schumacher has pointed out: "Modern man does not experience himself as a part of nature but as an outside force destined to dominate and conquer it" (1973: 13). The ecological movement has sought to promote the opposite idea: "Only insofar as the ecology movement consciously cultivates an anti-hierarchical and a non-domineering sensibility, structure, and strategy for social change can it retain its very identity as the voice for a new balance between humanity and nature and its goal for a truly ecological society" (Bookchin, cited in Dobson 1991: ix). In addition, Greens have argued that not only capitalists but also the historic socialist movements were proponents of industrialism and presumed that the best way to increase human welfare was to promote economic growth (Porritt 1985: 43–44).

Murray Bookchin (1991) proposed a comprehensive philosophy, which he called social ecology. He held that environmental dislocations were deeply rooted in a complex, irrational, and antiecological society. These problems, he argued, originated in a hierarchical, class, and competitive capitalist system that brings about a view that human beings are destined to dominate nature. This combination of an anarchist critique of hierarchy and exploitation with an ethic based on biological interdependence is meant to counter the emergence of New Age romanticism and the production of mystical ideologies, which appear under the names of deep ecology, Earth Goddess worship, and ecological animism. For Bookchin, ecology "has always meant social ecology: the conviction that the very concept of dominating nature stems from the domination of human by human, indeed, of women by men, of young by the elders, of one ethnic group by another, of society by state, of the individual by bureaucracy, as well as of one economic class by another or a colonized people by colonial power" (1989: 76). The ecological movement seeks to use science against science to show that industrial production, including science and technology, has been harmful to nature.

Ecofeminism began when feminists started to elaborate the similarities between male mistreatment of women and human mistreatment of nature in the 1970s.[5] Ecofeminists basically are feminist thinkers and activists who are concerned with connections between women and nature. Chris Cuomo succinctly formulates ecofeminism as "an umbrella term referring to forthright attempts to link some versions of feminism and environmentalism" (Cuomo 1998: 6). The term "ecofeminism" was invented by Françoise d'Eaubonne, who in 1974 invited women to lead an ecological revolution to save the planet (Merchant 1992: 184). According to Dobson, three main ideas form the foundation of ecofeminism. First, ecofeminists assert that there are certain intrinsic values and ways of behaving that reject the domination of nature and that are usually and more frequently possessed by women than by men. Second, the structures of the domination of women and the domination of nature are parallel and the motivations are the same. Third, they argue that women are closer to nature than men; therefore, women potentially become the vanguard in the development of sustainable ways of connecting to the environment (1992: 192–93).

Ecofeminism theorizes both the realities and the specificities of women's oppression, while explicitly giving attention as well to the exploitation of nonhuman beings and entities. This constitutes not only an expansive feminism but also a challenge to ecological and environmental thought, which ignored gender, race, class, or other connections between destructive ideologies and practices. Hence, ecofeminists put themselves at the crossroad of feminist, antiracist, and environmentalist movements and also critique capitalism, heterosexism, homophobia, and other features of oppression on the basis of the dualistic construction and maintenance of inferior, devalued, or pathologized "Others" (Cuomo 1998: 23–24). In *The Death of Nature* (1980), Carolyn Merchant describes how the scientific revolution effected the marginalization of women and nature, and how conceptualizations of nature by both science and religion significantly influenced the form and force of the human impact on nature and communities. Merchant traces how the descriptions and metaphors of mechanistic science were influenced by violence against women, especially the investigation of women accused of witchcraft throughout Europe during the early seventeenth century. "The exploitation of nature and animal is justified by feminizing them; the exploitation of women is justified by naturalizing them" (cited in Warren 1996: 37).

In her book, *Staying Alive: Women, Ecology, and Development*, Vandana Shiva asserts the utility of a "feminine principle" ecofeminism in empowering rural Indian women and enabling them to sustain their livelihood by interrupting deforestation (Shiva 1989; see also Cuomo 1998: 138–39). For Shiva, the "feminine principle," which includes "equality in diversity," has been destroyed by "Western male-oriented concepts and values," including the Western model of economic development, which Shiva and others have labeled "maldevelopment" (Shiva 1989).

Feminist political ecology tries to link feminist critiques of science (Shiva 1989; Mies & Shiva 1993; Merchant 1980, 1989; Haraway 1989, 1991; Harding 1986, 1987) and the analyses and actions of feminist and environmental movements, which are termed "fragmented thoughts," into a single perspective, which "treats gender as a critical variable in shaping resource access and control, interacting with class, caste, race, culture, and ethnicity to shape processes of ecological change, the struggle of men and women to sustain an ecologically viable livelihood, and the prospects of any community for 'sustainable development'" (Rocheleau, Thomas-Slater, & Wangari 1996: 4; see also Hart 1991; Ghai & Vivian 1992; Tsing 1993; Pankhurst 1992; West & Blumberg 1990).

Ecological feminism, as proposed by Cuomo, was born as "a cluster of perspectives that constitutes a subcategory of the Ecofeminist project, and which is noteworthy in its emphasis on the similarities and relationships between and among various forms of oppression, exploitation, and domination" (Cuomo 1998: 22). This strand of ecofeminism criticizes the mainstream environmental movements, deep ecology, and masculine ecological movements, but also offers strong criticism of ecofeminism itself.

[While some ecofeminists] tend to respond to cults of masculinity by holding femininity as the superior mode of being . . . ecological feminists argue that both women and nature are considered and constructed as feminine, that the inextricability of masculinity and femininity as concepts and as cultural products make it impossible to reclaim one without assuming the other, and that "femininity" is a potent tool for domination and control in general. . . . Ecological feminists believe that emphasizing the similarities between women and natural states or entities maintains a lack of attention to the ways in which men are natural beings, and women are also dominators and oppressors. (Cuomo 1998: 23)

She criticizes Arne Naess and other deep ecologists for their lack of concern for human interaction, the multiplicity and importance of human needs and goods, the complexities of human histories, and the relations between environmental issues and problems and human social and political reality, which she calls the "social" sources of ecological realities (Cuomo 1998: 77).

Ecological feminism, she says, is a philosophical feminism that tries to map out carefully its constituent concepts, logic, knowledge, and justifications and that stands on the shoulders of the wealth of feminist philosophical works of the last few decades. It locates "ecological feminism in the context of other works in ecology and social justice, and discusses the distinction—central to environmental ethics—between instrumental and non-instrumental values" (Cuomo 1998: 9). Ecological feminism is, thus, the most comprehensive challenge to the dominant production system, including Western knowledge and capitalist production.

The ecological movements have been institutionalized into the university system through academic journals and programs of study.[6] Most of the programs were established through the development of science departments such as forestry, natural resources, geography, and even engineering. But there are now courses "designed by historians, sociologists, architects, literary critics, and others. The range of institutional settings is no less diverse: from liberal arts college to schools of architecture, engineering, and agriculture" (Piasecki, 1984: 310).

Such institutionalization and diversification of environmental studies have not necessarily strengthened the ecological movement. In fact, in many ways they have diluted the basic objective of the movement by shifting the concern with ecological issues from the search for the origin of environmental problems to the management of the environment for production.[7] Environmental issues have been combined with science and presented as academic courses such as environmental chemistry, environmental geology, environmental economy, environmental management, environmental engineering, environmental archaeology, and so forth.

The ecological movement that emerged in the early 1960s has shifted from purely ecological or scientific concerns to philosophical ones, as in the deep ecology, social ecology, ecofeminist, and ecological feminist movements, all of which brought social science into the picture. They examined the responsibility of science in the environmental crisis and questioned human domination of nature as well as human domination over other human beings in the form of oppression on the basis of generation, class, gender, race, or indigeneity. On the other hand, the academic institutionalization of the ecology movement has in many ways diluted its political message, and the movement is increasingly co-opted by traditional disciplines, a partial revenge of the mainstream environmental movement over the ecological movement.

Notes

1. See Bramwell (1994), who distinguishes between reform ecologists and deep ecologists. Reform ecologists believe that you can work through and with the existing system. This group believes in working with industry and taking account of the need for economic growth. "Ecologism seeks radically to call into question a whole series of political, economic and social practices in a way that environmentalism does not. ... While most post-industrial futures revolve around high-growth, high-technology, expanding services, greater leisure, and satisfaction conceived in material terms, ecologism's post-industrial society questions growth and technology, and suggests that the Good Life will involve more work and fewer material objects. Fundamentally, ecologism takes seriously the universal condition of finite resources and asks what kinds of political, economic and social practices are (a) possible and (b) desirable within that framework" (Wall 1994: 205).

2. "*Silent Spring* and the controversy it produced brought science into the wide arena of public understanding and debate for the first time since the end of World War II. Carson convinced those who read her book that there was a fragile partnership between humans and nature, which once broken, could lead to the destruction of both. By providing an alternative vision of scientific progress, one that required an informed and vigilant citizenry, she launched a popular movement she never dreamed possible" (Lear 1993: 40).

3. "Ecology" originally came from *Oekologie*, a German word invented in 1866 by Ernest Haeckel. He defined ecology in his *General Morphology* as "the science of relations between organisms and their environment," which remains the core of the academic definition today (Allison 1991: 26).

4. "Ecology is now fashionable, indeed, faddish—and with this sleazy popularity has merged a new type of environmentalist hype. From an outlook and movement that at least held the promise of challenging hierarchy and domination has emerged a form of environmentalism that is based more on tinkering with existing institutions, social relations, technologies, and values than on changing them. I use the word 'environmentalism' to contrast it with ecology, specially with social ecology" (Dobson 1991: 60).

5. Elizabeth Dodson Gray's antihierarchical *Green Paradise Lost* (1981), Susan Griffin's *Women and Nature* (1978), and Carolyn Merchant's *Death of Nature* (1980) were published between 1978 and 1981; Léonie Caldecott and Stephanie Leland's collection, *Reclaim the Earth: Women Speak Out for Life on Earth,* appeared in 1983; and ecofeminist issues and theories were subsequently visible in various feminist and lesbian-feminist journals, periodicals, and books.

6. According to *Peterson's Guide to Four-Year Colleges* and *Peterson's Guide to Graduate and Professional Programs,* undergraduate programs of ecological studies are offered in more than 200 colleges and about 94 graduate schools in North America, as of 1999. A library search of ecology-related journals revealed that the journals published between 1913 and 1949 were all ecological science journals, with ecology

considered a branch of the biological sciences, under either world plant or world animal headings. The journals that were established in the 1960s tended to be broader in scope and more interdisciplinary than in the previous era. Most (political) ecological journals, such as the *Ecologist* (United Kingdom), *Ambio and Oikos* (Sweden), and *Environmental Review* and *Earth First!* (United States), emerged in the 1970s. In the 1980s and 1990s, they became still more diverse, both geographically and academically: *Environmental Design* (1985), *Environmental Studies* (1985), *Capitalism, Nature, Socialism* (1988), *Journal of Environmental Science* (1991), *International Journal of Environmental Ethics* (1993), *Democracy and Nature* (1995), *Journal of Environmental Planning* (1997), *Environmental Archeology* (1998), and *Global Ecology and Biogeography* (1999).

7. "Since 1984, there is more emphasis on the management of environmental solutions as opposed to the discovery of problems. . . .This emphasis on management also reflects a larger sociopolitical development that sets the context in which these classes meet. Environmental concerns are no longer perceived as separatist, the products of subversive sciences and disaffected scholarship" (Piasecki, 1992: 3).

*

13

The "Culture Wars" and the "Science Wars"

Richard E. Lee

The "culture wars" and the "science wars" that have pitted the defenders of truth, objectivity, reason, and Western civilization against the protagonists of secular values, relativism, and multiculturalism over the past two decades constitute more than just a simple, if particularly acrimonious, struggle between the old and the new, between the modern and the postmodern. They belong to the long history of debate and outright conflict in the modern world over how valid knowledge may be produced, what are the grounds and the domain of its authority, who may speak in its voice, and thus what courses of social action may be considered legitimate. In effect, this is the story of this book.

The major themes of the contemporary "culture wars" are not limited to the United States. Instances of debates over history and historical amnesia, language, education, and the "canon," including their gendered and racial/ethnic foundations and the power relations ensuring their reproduction, may be observed throughout the world. These "cultural" mechanisms have become centers of controversy because of their common functions, especially as techniques through which "nation" and "nationalism" have taken form as consequences of the construction of "otherness" and its attendant struggles, positioning groups in the hierarchy of the world division of labor.[1] As Gregory Jay notes: "The struggle for representation knows no borders. Many nations are trying to find a way to balance the claims of individuality, ethnic or racial solidarity, democracy, economic development, women's liberation, and nationalism" (1997: 62). For instance, salient issues in twentieth-century German history have been race and anti-Semitism, coming to terms with the Holocaust—uncomfortable terrain for German historians on both the right and the left until at least the 1960s (Herf 1995)—as well as current debates about "guest workers" and the definition of nation. Similar controversies have erupted in Austria with respect to the careers of such prominent political figures as Kurt Waldheim and Jörg Haider. In Japan, analogous questions have surfaced in discussions of Korean comfort women (Hicks 1994; Yoshiaki 2000) and the Nanjing massacre (Fogel 2000; Bix 2000). The contested nature of the relationship of the national state to its ethnic constituents has been especially pronounced in Africa, and an enormous literature has explored fundamentalism in the Arab world, especially Iran and Afghanistan, and the issues of representation and citizenship in the states coming out of the breakup of the Soviet Union, as well as in Israel and South Africa.

Nonetheless, it has been in the United States that these controversies have been grouped under a single rubric. Heated discussions associated with nationalism, patriotism, and the role of religion and with race, ethnic, and gender relations have developed around the same dilemmas over who gets what as are evidenced abroad. In the United States, however, there has been a clear articulation of the mechanisms relating putatively universal ideals and the particularist context in which they function. Furthermore, the range of conflicts associated with the culture wars in the United States, including the family, art, religion, education, law, and politics, is such that the culture wars touch "virtually all Americans," according to James Davison Hunter (1991: xi). Hunter's 1991 book, entitled simply *Culture Wars*, did much to popularize the term and crystalize the issues, which Hunter felt were quite different from the notorious late nineteenth-century *Kulturkampf* in Prussia. The latter merely sought to bring Catholic institutions under state (that is, Protestant) control, whereas today's culture wars, in the wake of the decline in the sectarian differences between Protestants, Catholics, and Jews, resulted in battle lines that unite persons of all three religious backgrounds on both sides of a debate defined as one between "orthodox" and "progressive" worldviews. As an illustration, for some domestic-partnership legislation represents an effort to redefine the family and constitutes a fundamental attack on religion. The issue is said to be broader than homosexuality; it is the demands of secular humanism, which represent, for these persons, challenges to received values. For others, it is simply an effort to give legal recognition to the reality of contemporary flexible and evolving units of social reproduction and communal arrangements.

In a similar vein, abortion may be viewed as a manifestation of moral decline or as an expression of freedom and justice, and the issue has implications for the debates over the allocation of responsibility, for instance, to the individual or to the larger social whole for access to health care. Opinions on such questions have often given rise to surprising coalitions.

All the parties in these and similar debates seem to have identified the issues as having a long-term component that speaks to the reproduction of social norms through education—what should be included in textbooks, how specific narratives of the past legitimate specific hierarchies in the present, what it might be possible to imagine for the future. All have understood that differing positions imply value-laden choices and thus are profoundly political in nature and speak to clear but contested power relations in a struggle for control, even though the debates may take the form of arguments over truth and relativism, tolerance and prejudice.

Indeed, if there is an area where "crisis," implicit in the war metaphor, has reached strikingly sharp focus, it is in the field of education. As the agendas directing values, morality, and acceptable conduct that are set at the top eventually redirect programs in elementary and secondary schools, it should be no surprise that debates have tended to implicate the content of higher education and the organizing and legitimating role of the university: as symbolic system (the "canon") and as material practice (access or gatekeeping, disciplinary/departmental structures, and pedagogical and research organization).

University reform, of which a significant aspect involved the curriculum, stands as an enduring legacy of the 1960s. As activism declined in the streets, a major component of the struggle was transferred to the classroom with its new mix of women and ethnic minorities and activist professors. The social sciences gave *droit de cité* to interpreting difference alongside the dominant mode of explaining variance within a universalist context; the new historicists began unearthing the anonymous world of past resistance and survival; and literary scholars opened up the historical dimension of the values that had informed interpretation and investigated just how culture accomplished its controlling function.

In one of their most widely recognized manifestations in the United States, the culture wars have been bound up with the struggle over teaching American literature, which has been subjected to new and often disconcerting interpretations, as well as demands for new inclusions and exclusions in the canon, or even the canon's total abolition. These demands to reopen the canon were closely related to issues of race and gender and were often linked explicitly to the movements of the 1960s (e.g., Hull, Scott & Smith 1982; Moraga & Anzaldúa 1983). Recovery (e.g., Fetterley 1985; Gates 1992) and classroom experimentation (e.g., Cain 1994; Alberti 1995) have been central in remaking the canon. Gregory

Jay has a point when he states that "it is precisely because higher education has done so much (though not enough) to redistribute access to representation that colleges and universities have come under such vitriolic attack" (1997: 57).

Come under attack they certainly did, in a whole series of gloom-and-doom writings.[2] Two works stand out: Allan Bloom, *Closing of the American Mind: How Higher Education Has Failed Democracy and Impoverished the Souls of Today's Students* (1987); and E. D. Hirsch, *Cultural Literacy: What Every American Needs to Know* (1988). These and their like presented an agenda for, and outlined the purpose of, shaping public schooling and higher education that "abstracted equity from excellence and cultural criticism from the discourse of social responsibility" (Giroux 1992: 123). As many critics immediately realized, Hirsch's culture was unproblematic because he had removed it from the dynamics of its construction in struggle and power. Barbara Herrnstein Smith states that, from a position of high institutional authority,

> *Cultural Literacy* (book, list, term, and concept) does a very good job of obscuring the nation's very real educational problems and assuring the American public, many of whom are naturally happy to hear it, that those problems (along with other social and economic ills . . .) are caused primarily by befuddled education professors and school administrators following what scientists have shown to be the incorrect principles of progressive education and, consequently, can be solved by school reforms that require no funding, entail no social or political changes, create no uncomfortable feelings for anyone except teachers and school administrators, and do not touch the structures of a single American institution, including its school system. (1992: 88–89)

It was in this context of the crusade to create "a narrowly specific cultural capital [as] the normative *referent* for everyone, but [to remain] the *property* of a small and powerful caste that is linguistically and ethnically unified" that the Stanford curriculum debate (and many such in universities all over the United States) was played out during the mid-1980s. Stanford had had a Western civilization course from 1935 to 1970,[3] and by 1980 it had reinstituted such a course that eventually included the Eurocentric and monumentalist reading list evincing the "structure of otherness" that, "depending on your perspective," either constitutes "the main *obstacle to* or the main *bulwark against* relational approaches to culture that sparked the debate" (Pratt 1992a: 15, 17, 19–20).

The culture wars and the parallel, anti–"political correctness" (PC) movement took place in a well-defined material context:

> In the present multinational, consumerist, postmodern corporate climate, socialist or populist critique is sidestepped or flat out banned as PC. Any talk of egalitarian principles is scorned or spurned as hope-

lessly PC. Further, in this context, the democratized university is an inefficient, obsolete notion and not cost-effective, so the PC scare gives a rationale for dismantling it, or at least as it is currently constituted. (J. Williams 1995: 5)

Nonetheless, PC attests to the force of the social critiques emanating from the humanities and the social sciences, and the anti-PC movement shows to what extent challenges have had an effect. From the 1960s, colleges and universities, especially public institutions, were pressured to serve greater and greater numbers of students. This expansion, however, took place during a worldwide economic downturn, which resulted in an overall contraction of resources, even though the extent of the shortfall was politically determined. Thus it has been suggested that, by attacking PC, "conservatives have implemented a well-orchestrated and financed campaign to cut budgets, downsize universities, and thus sharply restrict access to higher education." Given the generalized sense of entitlement to education, access can now "*only* be restricted if one can successfully argue that restriction is a function of economic forces beyond our control, and if one can somehow make colleges politically suspect" (Lauter 1995: 73, 81).

The arguments over the politically suspect politicization of the university that ranged those who viewed the realm of knowledge and its institutions as loci of struggle against those who upheld "truth" achieved through value-neutrality as the ideal of scholarship seem paradoxical in the light of actual practice. As Gerald Graff contends:

> It is the academic left that has arguably made the major contributions to scholarship in literary studies over the past three decades, profoundly affecting fields as remote from literature as law, architecture, anthropology, and the social sciences. Yet it is the right that defends the virtues of disinterested scholarship, while the left debunks disinterestedness in the name of politics and power. The producers of the best objective scholarship defend partisanship, while the defenders of objectivity produce mostly partisan political polemics. (1995: 308)

The paradox is further compounded when the trajectories of "theory" and "identity politics" are considered.

Theory, whether loosely characterized as post-structuralism or deconstruction, unmasked the power relations inherent in the Arnoldian high-culture aesthetic. It was anathema on the right but also tended to result in formal analyses of "texts," of whatever type, disengaged from concrete political agendas and was roundly criticized on the left. According to Joan Scott: "On the right, there are denunciations of the nihilism of theory, which, it is said, will leave us orphans without cultural patrimony. On the left, theory is indicted for its impracticality: it does not connect to 'real life' or 'lived experience' and so cannot lead directly to politics, to revolution, or at least to social reform." Scott goes on to point out that "[i]n the attack on 'theory,' right and left clear the field of all possible critiques of their foundational premises; with those intact, they can fight safely and familiarly among themselves" (1995: 301).

Unlike the antiessentialism of post-structuralism, identity politics presupposes foundations making the point that without presuppositions there can be no action on the part of the marginalized. For Stuart Hall, the poor fit between theory and practice opens up an opportunity for rethinking the terms of the debate. Identity politics, he argues, even in the form of so-called strategic essentialism,

> sees difference as "their traditions versus ours," not in a positional way, but in a mutually exclusive, autonomous and self-sufficient one. And it is therefore unable to grasp the dialogic strategies and hybrid forms essential to the diaspora aesthetic. A movement beyond this essentialism is not an aesthetic or critical strategy without a cultural politics, without a marking of difference. It is not simply rearticulation and re-appropriation for the sake of it. What it evades is the essentializing of difference into two mutually opposed either/ors . . . replac[ing] the "or" with the potentiality or the possibility of an "and." That is the logic of coupling rather than the logic of a binary opposition. . . . The essentializing moment is weak because it naturalizes and dehistoricizes difference, mistaking what is historical and cultural for what is natural, biological, and genetic. (S. Hall 1996: 472)

Who controls that specific representational form we call "history" and to what end has been a matter of serious controversy over the past decade (e.g., Linenthal & Englehardt 1996; Windschuttle 1997; Wood & Foster 1997), although one must acknowledge again that this is not new, and the university is again at the center of struggles over what, or whose, history is recognized as authoritative and propagated as grounds for legitimate social action. Indeed, the curriculum is a historically specific narrative and pedagogy a particular form of cultural politics.

However, when this historical dimension is taken into consideration, the contemporary culture wars appear as a post-1968 conjunctural moment (Lee 1996) in a long-term trend (Dejean 1997). Neither the intellectual arrangements nor the institutional organization of the structures of knowledge was able to negotiate successfully the outcomes of the challenges of the 1960s. And now, once students graduate, it is apparent that the promise of a better life for all who go through the system cannot be kept in a world ruled by the law of value. Ira Shor (1992) has traced how the culture wars have been bound up with education politics as the curriculum was manipulated to intervene in global crises highlighted in the struggles of the 1960s: First, in the search for reform and the restoration of order, the focus was directed to vocational and career-oriented training in the early 1970s. Then the "literacy crisis" spawned the "back-to-the-basics" movements of the late 1970s through the early 1980s. And, finally, the mid-1980s saw the beginning of a two-pronged drive for "austerity" plus "excellence" and educational discipline. Shor notes the striking reversal from the first period to the third, and argues that "[w]hile careerism offered too little to inspire students, the new academic mystique promised everyone too much" (1992: 163).

In an article entitled "The Culture War That Isn't," published in the Heritage Foundation's *Policy Review*, Jeremy Rabkin notes that "[w]hat is perhaps most striking about the prevailing defeatism among religious conservatives is its ahistorical, perhaps even anti-historical, character. It's as if the collision of religion and politics only happened yesterday" (1999: 8). But as Hall's analysis suggests, the outcome of the contemporary controversies may have profound consequences for the dichotomous thinking, and particularly the divorce of facts from values, that has formed the basis of the epistemology, the geoculture of the modern world-system as it has developed over the past five centuries (Lee 2000; Lee & Wallerstein 2001). In this case it is "civilization-culture" itself—the first universal, implicit in the term as it was deployed by Lynne Cheney (1988), and the second, particularist—that is in question.

More to the point, at issue today is the epistemological basis of the representational apparatus that has characterized our understanding of the social world over the long term.

> What the political correctness debate and related phenomena display, symptomatically, is precisely the connection between representation in the field of knowledge and representation in the fields of society and politics. . . . [It is the impression that] academics are producing *a body of different truths* that threaten certain traditional value systems and institutions [that has engendered the backlash of the culture wars]. (Jay 1997: 31)

Far from the "orthodox" versus "progressive" elements that Hunter postulated, the protagonists of the culture wars simply cannot be shoehorned into "either/or" groups, of whatever stripe. Moreover, in this sense, Hall's analysis suggests that the conjunctural moment of the culture wars is also a point of structural crisis leading to a restructuring in the field of knowledge—the delegitimation of the principle of the excluded middle and the implicit call for a relational conception of human reality. But the epistemological status of dichotomous thinking is constitutive of and constituted by the role of the natural sciences in the modern world and their location at the privileged pole of the structures of knowledge that has become the object of contention in the science wars.

At first blush the "science wars" may seem to be nothing more than a specifically focused instance of the more generalized "culture wars." Indeed the two do share several characteristics. First, the themes of the debates constituting the science wars, like those of the culture wars, have a long history. Second, as the institutions of knowledge formation were opened up to previously excluded groups as a result of the upheavals of the 1960s, entrenched theoretical and methodological perspectives found it more and more difficult to dismiss the critiques and their overtly political agendas that were now mounted from within the institutions themselves. It became clear to many involved that any review of the premises of knowledge formation amounted to an evaluation of the structure of social relations as well. This dimension of the culture wars was replicated in the science wars. Finally, the disputes in both arenas, if indeed there are two, were polarized and the tone exacerbated by a series of publications defending what had been, up to the 1960s, the dominant positions.

At the end of the seventeenth century, just as the questions of universalism and particularism were being debated in both the French *querelles des anciens et des modernes* and the English Battle of the Books in what we would now consider the humanities, Isaac Newton, as we have seen, was working out a rationalist model, combining mathematical representation and empirical observation, for the analysis of physical reality based on universal natural laws. It linked the particular facts that were the product of observation through and with the universal facts of time-reversible equations. The solution allowed for prediction and would eventually succeed thereby in jettisoning value-laden, thus relativist, explanatory modes. Let us remember also that by the turn of the nineteenth century, Laplace could state that he did not need the hypothesis of God to arrive at his proofs. Nonetheless, the adequacy of science *tout court*—its pretensions of truth and objectivity and their association with progress—remained an object of controversy: the Tory Radical opposition to industrial society and the "dismal science" during the first half of the nineteenth century; the exchange between the liberal-humanist critic Matthew Arnold and the apologist for the physical sciences T. H. Huxley in the second half; and a similar dispute some four decades ago between the literary critic F. R. Leavis and the physicist C. P. Snow in the mid-twentieth century. What does seem clear is that the "general esteem for this relatively new hybrid—science-technology—seems to have peaked in around 1960" (Trachtman & Perrucci 2000: 8–9).

The chronology of the science wars proper may well be taken to begin in 1992, with the publication of the books by physicist Steven Weinberg, *Dreams of a Final Theory: The Search for the Fundamental Laws of Nature* (1992), and by biologist Lewis Wolpert, *The Unnatural Nature of Science: Why Science Does Not Make (Common) Sense* (1993). These two books brought to public attention the work of the group of historians, philosophers, and sociologists who had been involved in reconceptualizing many common views about the nature of science (Wolpert 1993: 110–11, 115–17; Weinberg 1992: 184–90). For Wolpert, the bone of contention is relativism, about which he replies that "[s]cientists can be very proud to be naive realists" (1993: 117). Weinberg agrees but contends that relativism "is only one aspect of a wider, radical, attack on science itself." He goes on to give examples from Paul Feyerabend, Sandra Harding, and Theodore Roszac and echoes one of the themes of the culture wars in suspecting that "Gerald Holton is close to the truth in seeing the radical attack on science as one symptom of a broader hostility to Western civilization" (1992: 189–90). It should be noted that these attacks were directed primarily at the strongest social constructivist and relativist bent of science studies. The fact that these views were controversial even within science studies was, and still is, often overlooked.

In any case, by the early 1990s it had become "apparent that some scientists felt sufficiently threatened that they were impelled to go public with a defense of the rationality and the benevolence of science and an attack on what they viewed as uninformed, biased, and unwarranted criticism" (Trachtman & Perrucci 2000: 24). The issues came to a head with the "defense" mounted on a wide front in 1994 by the publication of *Higher Superstition: The Academic Left and Its Quarrels with Science*, by Paul Gross, a biologist at the University of Virginia, and Norman Levitt, a mathematician at Rutgers University. Their scattershot attack was directed at a broad "academic left"—feminist theory, postmodern philosophy, deconstruction, deep ecology—and the way it "dislikes science."And these dislikes extend not just to the uses to which science is put, but also to

> the social structures through which science is institutionalized, to the system of education by which professional scientists are produced, and to a mentality that is taken, rightly or wrongly, as characteristic of scientists. Most surprisingly, there is open hostility toward the *actual content* of scientific knowledge and toward the assumption, which one might have supposed universal among educated people, that scientific knowledge is reasonably reliable and rests on a sound methodology. (Gross & Levitt 1994: 6, 2)

The authors touched a chord in the scientific community despite the fact that the book itself stands as a monument to the very type of intellectual dilettantism it attacks. In an analysis of the "reading" of the work of Jacques Derrida by protagonists of the culture wars, Arkady Plotnitsky (who has both scientific and literary credentials) writes that

> [i]n general, scholarly problems of monumental proportions are, to use the language of topology, found in the immediate vicinity of just about every point of *Higher Superstition*. It is not so much embarrassing errors even as egregious as that of the misreading of *"topique différantielle"* as differential topology, that are most crucial (we all make mistakes, sometimes absurd mistakes), but the intellectually and scholarly inadmissible practices and attitudes that pervade—and *define*—

this sadly irresponsible book. Gross and Levitt's warning concerning "threats to the essential grace and comity of scholarship and the academic life" (ix) becomes, in one of many bizarre ironies of the book, its self-description. . . . The tragedy is that so many scientists, including some among the best scientists, have taken it seriously and accepted its arguments, and even adopted its unacceptable attitudes. (1997: n.p.; emphasis in original)

The exchanges that ensued made plain just how fundamental the issues were, and not simply that Snow's "two cultures" of scientists and nonscientists could still not communicate with one another.

The National Association of Scholars joined in the responses to the supposed "antiscience" threat (see Holton 1993 on the dangers of irrationalism) with two conferences arranged in 1994. The "Science in American Life" exhibit arranged by the American Chemical Society, which incorporated criticism of science, prompted an exchange between Tom Gieryn (1996) and Paul Gross (1996). Harry Collins and Lewis Wolpert confronted one another directly at the British Association for the Advancement of Science in 1994, and Fuller organized a follow-up to this conference that sought some mutual understanding but ended with disappointing results (see Fuller 2000). Gross and Levitt themselves proved unforthcoming at the debate on the content of their book at the Society for Social Studies of Science annual meeting in 1995. The meeting was punctuated, however, with an unseemly exchange over the scientific credentials of Donna Haraway and other feminist critics. In 1995, the New York Academy of Sciences sponsored a conference entitled "The Flight from Science and Reason," which engaged the questions raised in *Higher Superstition* by a long list of defenders of science against such "science bashers" as constructivists, deconstructionists, creationists, feminists, Afrocentrists, and radical environmentalists ("ecosentimentalists") (Gross, Levitt, & Lewis 1996).

What was missing was a concerted response to Gross and Levitt, and their "pro-science" supporters. The editors of *Social Text* took on this project in 1996 with a special issue dedicated to the science wars. What the editors of the journal did not know was that the physicist Alan Sokal, inspired by his reading of *Higher Superstition*, was involved in an "actively sustained" conspiracy to deceive the journal into publishing his "Transgressing the Boundaries: Toward a Transformative Hermeneutics of Quantum Gravity" in which he "parodied postmodern stylistic conventions and derived politically correct conclusions from an esoteric subfield of science" (Segerstråle 2000: n.p.).[4] Sokal exposed the hoax in "A Physicist Experiments with Cultural Studies," which appeared almost concurrently in *Lingua Franca*, characterizing the article as combining "nonsense" and "silliness." For those who were taken in by Sokal and those whom they represented, it was an extraordinary breach of intellectual ethics and scholarly integrity. For those making common cause with Sokal (and Gross & Levitt et al.), it forcefully demonstrated his point about declining "standards of rigor in the academic community" (Weinberg 1996: 11) and specifically the intellectual laxness of those it was designed to attack. This group proved notoriously difficult to define, however, given the diversity of the cultural studies community and the conflation of cultural studies with science studies as the debates played out.

The defenders of traditional science were concerned to protect what they considered to be the "true" idea of science against what they described as "false" representations. For their critics, these defenders seemed not "to recognize the right of other academics to do their own interpretations of science within the particular frameworks of their own disciplines" (Segerstråle 2000a: 21). The problem for the critics of traditional science was that the latter's views permeated far beyond the confines of their specialist disciplines. They had become part both of the undergraduate curricula and of the larger public understanding of science. Ullica Segerstråle relates this to an internal struggle within the political left (recalling the culture wars): the older left of Gross, Levitt, and Sokal "equates science with reliable knowledge, a tool in the struggle for social justice. . . . In contrast, the cultural Left equates science with power that can be used for social oppression of minorities. For them, therefore, science criticism is a way to liberation." The issue, they feel, is whether nonscientists would "have some say in the decision-making process of the professional scientific community" (Segerstråle 2000b: n.p.).

Steve Fuller addresses this point directly, contending that instead of the generalized condemnation of higher education that was one of the results of the Sokal affair,

> a more productive debate would realign the parties so that scientists and STSers who wish to protect the academy from the rest of society could stand on one side, while those who wish to use the academy as a vehicle for reforming society could stand on the other, . . . [T]his repositioned debate would not reproduce natural (disciplinary) divisions within the academy but would force academics to seek constituencies outside academia for whom alternative conceptions of the social role of academics could make a difference to their own activities. (Fuller 2000: 209)

Unfortunately, the fallout from the Sokal affair has obscured Andrew Ross's own take on the science wars. In his introduction to the *Social Text* collection, he treats the science wars as a "second front" in the "holy Culture Wars" (1996a: 6) and suggests a conjunctural explanation for the controversy, especially near-term decline in governmental support, including funding cuts, for big science (cancellation of the

Superconducting Supercollider and closing of the Congressional Office of Technology Assessment) that others have noted. However, this is not just a turf battle over the allocation of scarce research dollars, and Ross's analysis also presents a longer-term interpretation of the science wars that situates the only real resolution in a shift in the process privileging formal rationality to one favoring substantive concerns:

> The rise in technoskepticism, then, parallels a crisis in industrialization which is often mistaken for a crisis of the environment. . . . [T]he remoteness of scientific knowledge from the social and physical environments in which it will come to be measured and utilized is as irrational as anything we might imagine, and downright hazardous when it involves materials that can only be properly tested in the open environment. . . . [D]emonstrating the socially constructed nature of the scientist's knowledge . . . may help to demystify, but it must be joined by insistence on methodological reform—to involve the local experience of users in the research process from the outset and to ensure that the process is shaped less by a manufacturer's interests than by the needs of communities affected by the product. This is the way that leads from cultural relativism to social rationality. (A. Ross 1996a: 2, 3–4)

Ross also sets the "relativism" argument in a material, spatial, context and links it positively with "diversity": "Once it is acknowledged that the West does not have a monopoly on all the good scientific ideas in the world, or that reason, divorced from value, is not everywhere and always a productive human principle, then we should expect to see some self-modification of the universalist claims maintained on behalf of empirical rationality. . . . This is the way that leads from relativism to diversity" (Ross 1996a: 4). Seen together, Fuller and Ross reinforce Segerstråle's argument that most "constructivist sociologists (unlike their postmodern or cultural studies colleagues) are not primarily interested in values and ideology; they see themselves as epistemological radicals" (Segerstråle 2000b: n.p.). Indeed, as Harding insists: "There is plenty of science still to be done once physics is invited and permitted to step down and take its place as one human social activity among many others. What kinds of knowledge about the empirical world do we need in order to live at all, and to live more reasonably with each other on this planet from this moment on? Who should make up the 'we' who answers this question?" (1992: 20).

The science wars are not, however, simply a special case of the culture wars and a plea for more democracy and value-consideration in decision making in the context of the innate inequalities that are part of the all-encompassing corporate climate. Rather they indicate the importance of the questions focusing on the epistemological status of the sciences that have come into play. This becomes clear when we realize that what the science warriors seem not to have adequately considered is that "epistemological radicalism" is now found within the science community itself, especially among those involved in complexity studies.[5] The combination of the conviction that there is a "real" world and that the future, although it is "determined" by the past, is nonetheless unpredictable and the parallel assaults on dualism (e.g., Barrow 1995; Prigogine 1996) challenges the epistemological status of the sciences as discoverers, guardians, and purveyors of authoritative knowledge, that is, truth, by redefining what it means to describe the evolution of natural systems.

In sum, these developments underscore the covert, long-term, structural nature of the debates that have come to be known as the culture wars and the science wars. The consequences suggest an annulment of the presumed contradiction between determinism and free will, impinge directly on the manner in which scholars make claims for the legitimacy of their interpretations of social reality, and thus amount to overturning the dominant model shaping our understanding of the human world.

Notes

1. The war metaphor has also been applied to particular manifestations of the debate, e.g., the "history wars" and the "sex wars." In their work in this latter area, Duggan and Hunter say they seek to avoid "the twin dangers of narrow identity-based, single-issue politics on the one hand, and universalizing utopian projects on the other, [by turning] to a politics sensitive to specific local and historical contexts" (1995: 4).

2. They are legion; e.g., Bromwich (1992), Bloom (1987), Hirsch (1988), Boyer (1987), Wilshire (1989), Anderson (1992), Bennett (1984), Cheney (1988, 1990), Curtis et al. (1985), D'Souza (1991), Kimball (1990), Sykes (1988), Schlesinger (1992), Bernstein (1994), Sommers (1994). That of Bennett was especially important in that it was issued in his official capacity as director of the National Endowment for the Humanities. "Even Michael Lind's call for a renewed 'liberal nationalism' in *The Next American Nation* succumbs to anti-academic rhetoric, as well as echoing the neoconservative line that "'multiculturalism is not the wave of the future, but an aftershock of the black-power radicalism of the sixties' [and] Todd Gitlin, a real tenured radical, writes *Twilight of Common Dreams* to lament the 'breakdown of the idea of a common Left' and the rise of identity politics, which he blames on black separatists and their imitators who have unjustly demonized white male liberals" (Jay 1997: 3).

3. The background of the Stanford course included the Contemporary Civilization course instituted at Columbia in 1919, a follow-on to the War Issues courses of 1918 (including that of Columbia), instructing future conscripts in the European heritage they were called upon to defend (Pratt 1992a: 14; see also Allardyce 1982 and Lindenberger 1989).

4. The original article, the *Lingua Franca* revelation, responses, reactions in the press (domestic and foreign), and longer essays and colloquies on the "affair" are conveniently collected in Editors of *Lingua Franca* (2000); A. Ross (1996b) is an expanded version of the special issue of *Social Text* without the Sokal piece.

5. Ilya Prigogine has argued that the "sciences are not the reflection of a static rationality to be resisted or submitted to; they are furthering understanding in the same way as are human activities taken as a whole" (1988: 3). He goes so far as to state that "I believe that what we do today depends on our image of the future, rather than the future depending on what we do today" (Prigogine in Snell & Yevtushenko 1992: 28).

14

Conclusion?

Immanuel Wallerstein

What can we conclude from this analysis of the two cultures? We have tried to show how the idea that there are two cultures that should undergird our structures of knowledge was a creation of the modern world. We have tried to describe how this concept was gradually institutionalized in our universities and other intellectual institutions. We have emphasized the slow but definitive triumph of the sciences over the humanities in the search for public recognition and support, forming a hierarchy of intellectual worth in the eyes of the historical social system of which we are all a part, the capitalist world-economy.

This hierarchy of knowledge reached its acme in the period following 1945. Science seemed to offer untold technological advances. Scientific endeavors were disproportionately located in the United States, the dominant power in the world-system, and one that was ready to invest large sums of money furthering science. For one thing, scientific achievements seemed necessary to maintain the economic and military superiority of the United States. The competition with the USSR, known as the Cold War, maintained the pressure on the government to support the scientists. This pattern of support for science was replicated elsewhere, if on a somewhat smaller scale—in the Soviet Union and its allies, in western Europe, in Japan, and to a much lesser extent in the rest of the world.

Then came the earthshaking, worldwide revolutions of 1968. The consequences of these challenges to existing orders were contained in the political arena within several years, or rather their consequences in the political arena have been much slower to realize themselves. But these revolutions created an unceasing storm within the structures of knowledge. It has not been the same since. We have tried in this book to show the many turbulences bred by this cultural tornado. We have tried to argue that eventually all those who have been challenging the ways in which knowledge was structured in the modern world have arrived sooner or later at the fundamental epistemological question: are there two cultures, or one?

We do not pretend that the answers have been definitive. Nor that there is some new consensus in the world of knowledge. Far from it! We claim, however, that the epistemological debate has become central to the pursuit of knowledge in all of the traditional disciplines that constitute today the structures of knowledge. And we claim that this will continue to be so for at least the next twenty to thirty years.

At the end of this time of intellectual turmoil, for that is what it is, we shall perhaps emerge with a reunified epistemology. We hope so, although we are by no means claiming that it is inevitable that this happen. Nor do we pretend to spell out the details of what this reunified epistemology underpinning a set of reunified structures of knowledge would look like. But either we shall try to have an intelligent worldwide discussion about this, or not.

And if we do, it seems clear that the challenges raised by the new knowledge movements after the 1960s are serious and will have to be taken into account. These challenges have often presented themselves in exaggerated and polemical form. This is natural, but after a while the usefulness of such exaggeration is limited. There is serious work before us, and we must take it seriously. The chaotic disorder in the world of knowledge is part of a larger chaotic disorder of the world-system as a whole (Hopkins & Wallerstein, et al. 1996). The resolution of the epistemological issues will both affect and depend upon the resolution of the larger issues of the kind of historical social system we shall be building in the next thirty to fifty years. In this book, we have tried to present the issues that need to be resolved. The task of resolving them, however, is one in which we must all participate.

References

Abdel-Malek, Anouar. 1963. Orientalism in Crisis. *Diogenes* 44:107–8.

———. [1972] 1981. *Civilizations and Social Theory*. Vol. 1 of *Social Dialectics*. Translated by Mike Gonzalez. Albany: State University of New York Press.

Achugar, Hugo. 1994. *La Biblioteca en ruinas: Reflexiones culturales desde la periferia*. Montevideo, Uruguay: Ediciones Trilce.

Acuna, Rodolfo F. 1972. *Occupied America: The Chicano Struggle for Liberation*. San Francisco: Canfield.

Adas, Michael. 1989. *Machines as the Measure of Men: Science, Technology, and Ideologies of Western Dominance*. Ithaca, N.Y.: Cornell University Press.

Aida, S., et al. 1985. *The Science and Praxis of Complexity: Contributions to the Symposium Held at Montpellier, France, 9–11 May 1984*. Tokyo: United Nations University.

Alarcon, Norma. 1991. The Theoretical Subject of *This Bridge Called My Back* and Anglo-American Feminism. In *Criticism in the Borderlands: Studies in Chicano Literature, Culture, and Ideology*, edited by Hectór Calderón and José David Saldivar, 28–39. Durham, N.C.: Duke University Press.

Alberti, John, ed. 1995. *The Canon in the Classroom: The Pedagogical Implications of Canon Revision in American Literature*. New York: Garland.

Alborn, Timothy. 1996. The Business of Induction: Industry and Genius in the Language of British Scientific Reform, 1820–1840. *History of Science* 34, no. 103 (Mar.): 91–121.

Alcoff, Linda, and Elizabeth Potter. 1993. *Feminist Epistemologies*. New York: Routledge.

Allardyce, Gilbert. 1982. The Rise and Fall of the Western Civilization Course. *American Historical Review* 87:695–743.

Allison, Lincoln. 1991. *Ecology and Utility: The Philosophical Dilemmas of Planetary Management*. Teaneck, N.J.: Fairleigh Dickinson University Press.

Alten, Michèle. 1996. Référence révolutionnaire et chant scolaire sous la IIIe République (1880–1939). *Revue du Nord* 78, no. 317 (Oct.–Dec.): 975–86.

Anderson, Bonnie S., and Judith P. Zinsser. 1990. *A History of Their Own: Women in Europe from Prehistory to the Present*. 2 vols. Harmondsworth, UK: Penguin.

Anderson, Martin. 1992. *Imposters in the Temple: The Decline of the American University*. New York: Simon & Schuster.

Ankersmit, F. R. 1995. Historicism: An Attempt at Synthesis. *History and Theory* 34, no. 3:143–61.

Anzaldúa, Gloria. 1987. *Borderlands/La frontera: The New Mestiza.* San Francisco: Spinsters.

Aparicio, Frances, and Susana Chavez-Silverman, eds. 1997. *Tropicalizations: Transcultural Representations of Latinidad.* Hanover, N.H.: Dartmouth University Press.

Appadurai, Arjun. 1996. *Modernity at Large: Cultural Dimensions of Globalization.* Minneapolis: University of Minnesota Press.

Arico, José, ed. 1980. *Mariátegui y los origenes del marxismo latinoamericano.* Mexico City: Cuadernos Pasado y Presente.

Aronowitz, Stanley. 1993. *Roll Over Beethoven: The Return of Cultural Strife.* Hanover, N.H.: Wesleyan University Press.

Arrighi, Giovanni. 1994. *The Long Twentieth Century: Money, Power, and the Origins of Our Times.* London: Verso.

Arrighi, Giovanni, Terence K. Hopkins, and Immanuel Wallerstein. 1989. *Antisystemic Movements.* London: Verso.

Ash, Mitchell G. 1980. Academic Politics in the History of Science: Experimental Psychology in Germany, 1879–1941. *Central European History* 13, no. 3:255–79.

Ashby, Eric. 1947. *Scientist in Russia.* Harmondsworth, UK: Pelican.

Atlan, Henri, et al. 1985. *La sfida della complessità.* A cura di Gianluca Bocchi e Mauro Ceruti. Milano: Feltrinelli.

Auchmuty, Rosemary. 1996. Lesbian Studies: Politics of Lifestyle? In *All the Rage: Reasserting Radical Lesbian Feminism,* edited by Lynne Harne and Elaine Miller, 200–213. New York: Teachers College Press.

Ayer, Alfred Jules, ed. 1959. *Logical Positivism.* Glencoe, Ill.: Free Press.

Bacon, Alan. 1986. English Literature Becomes a University Subject: King's College, London as Pioneer. *Victorian Studies* 29, no. 4 (Summer): 591–612.

———. 1998. *The Nineteenth-Century History of English Studies.* Ashgate, UK: Aldershot.

Bacon, Francis. [1620] 1899. *Advancement of Learning and Novum Organum,* edited by James Edward Creighton. New York: Colonial Press.

Baker, Keith Michael. 1969. The Early History of the Term "Social Science." *Annales of Science* 20, no. 3 (Sept.): 211–26.

Bakhtin, Mikhail. 1984. *Rabelais and His World.* Bloomington: Indiana University Press.

Baldick, Chris. 1983. *The Social Mission of English Criticism, 1848–1932.* Oxford: Clarendon Press.

Bambara, Toni Cade. 1970. *The Black Woman: An Anthology.* New York: Penguin.

Barbour, Ian. 1997. *Religion and Science: Historical and Contemporary Issues.* San Francisco: HarperCollins.

Barkai, Haim. 1996. The Methodenstreit and the Emergence of Mathematical Economics. *Eastern Economic Journal* 22, no. 1 (Winter): 1–19.

Barker, Martin, and Anne Beezer. 1992. Introduction: What's in a Text? In *Reading into Cultural Studies,* edited by Martin Barker and Anne Breezer, 1–20. London: Routledge.

Barlow, Tani E. 1997. Colonialism's Career in Postwar China Studies. In *Formations of Colonial Modernity in East Asia,* edited by Tani E. Barlow, 373–412. Durham, N.C.: Duke University Press.

Barnes, Barry. 1974. *Scientific Knowledge and Sociological Theory.* London: Routledge & Kegan Paul.

Barrera, Mario. 1979. *Race and Class in the Southwest: A Theory of Racial Inequality.* Notre Dame, Ind.: University of Notre Dame Press.

Barrow, John D. 1995. *The Artful Universe: The Cosmic Source of Human Creativity.* Boston: Back Bay.

Barry, Peter. 1995. *Beginning Theory: An Introduction to Literary and Cultural Theory.* Manchester, UK: Manchester University Press.

Basalla, George. 1968. *The Rise of Science: External or Internal Factors?* Lexington, Ky.: Heath.

Bauman, Zygmunt. 1999. *Culture as Praxis.* Newberry Park, Calif.: Sage.

Beard, Mary Ritter. 1946. *Woman as Force in History: A Study in Tradition and Realities.* New York: Macmillan.

Bektas, M. Yakup, and Maurice Crosland. 1992. The Copley Medal (1731–1839). *Notes and Records of the Royal Society* 46, no. 1:43–76.

Bell, Matthew. 1994. *Goethe's Naturalistic Anthropology: Man and Other Plants.* Oxford: Clarendon Press.

Bell-Villada, Gene. 1996. *Art for Art's Sake and Literary Life.* Lincoln, Neb.: University of Nebraska Press.

Belnap, Jeffrey, and Raul Fernandez, eds. 1998. *Jose Marti's "Our America": From National to Hemispheric Cultural Studies.* Durham, N.C.: Duke University Press.

Ben-David, Joseph. 1971. *The Scientist's Role in Society: A Comparative Study.* Englewood Cliffs, N.J.: Prentice Hall.

———. 1977. *Centers of Learning: Britain, France, Germany, United States.* New York: McGraw-Hill.

Ben-David, Joseph, and Randall Collins. 1966. Social Factors in the Origins of a New Science: The Case of Psychology. *American Sociological Review* 31, no. 4:451–65.

Bendix, Rheinhard. 1956. *Word and Authority in Industry: Ideologies of Management in the Course of Industrialization.* Berkeley and Los Angeles: University of California Press.

Bennett, William. 1984. *To Reclaim a Legacy: A Report on the Humanities in Higher Education.* Washington, D.C.: National Endowment for the Humanities.

Benson, Donald R. [1981] 1985. Facts and Constructs: Victorian Humanists and Scientific Theorists on Scientific Knowledge. In *Victorian Science and Victorian Values: Literary Perspectives,* edited by James G. Paradis and Thomas Postlewait, 299–318. New Brunswick, N.J.: Rutgers University Press.

Bergson, Henri. 1922. *Durée et simultanéité: A propos de la théorie d'Einstein.* Paris: Alcan.

Berman, Morris. 1975. "Hegemony" and the Amateur Tradition in British Sciences. *Journal of Social History* 8 (Winter): 30–50.

Bernal, J. D. 1953. *Science and Industry in the Nineteenth Century.* London: Routledge & Kegan Paul.

———. [1954] 1971. *Science in History.* Vol. 2, *The Scientific and Industrial Revolutions.* Cambridge: MIT Press.

Bernal, Martin. 1987. *Black Athena: The Afroasiatic Roots of Classical Civilization.* Vol. 1, *The Fabrication of Ancient Greece 1785–1985.* Princeton, N.J.: Rutgers University Press.

Bernstein, Richard. 1994. *Dictatorship of Virtue: Multiculturalism and the Battle for the America's Future.* New York: Alfred Knopf.

Berr, Henri. 1899. *L'Avenir de la philosophie: Esquisse d'une synthèse des connaissances fondée sur l'histoire.* Paris: Hachette.

Bevilacqua, F. 1994. The Emergence of Theoretical Physics. In *Natural Sciences and Human Thought,* edited by Robert Zwilling, 13–36. New York: Springer Verlag.

Biagioli, Mario. 1992. Scientific Revolution, Social Bricolage, and Etiquette. In *The Scientific Revolution in National Context,* edited by Roy Porter and Mikulás Teich, 11–54. Cambridge: Cambridge University Press.

———, ed. 1999. *The Science Studies Reader.* New York: Routledge.

Bix, Herbert P. 2000. *Hirohito and the Making of Modern Japan.* New York: HarperCollins.

Black, Joel. 1990. Newtonian Mechanics and the Romantic Rebellion. In *Beyond the Two Cultures: Essays on Science, Technology, and Literature*, edited by Joseph W. Slade and Judith Yaross Lee, 131–39. Ames: Iowa State University Press.

Blauner, Robert. 1972. *Racial Oppression in America*. New York: Harper & Row.

Bleier, Ruth. 1984. *Science and Gender: A Critique of Biology and Its Theories on Women*. New York: Pergamon.

Bloom, Allan. 1987. *The Closing of the American Mind: How Higher Education Has Failed Democracy and Impoverished the Souls of Today's Students*. New York: Simon & Schuster.

Bloor, David. [1976] 1991. *Knowledge and Social Imagery*. 2nd ed. Chicago: University of Chicago Press.

Blundell, Valda, John Shepherd, and Ian R. Taylor, eds. 1993. *Relocating Cultural Studies: Developments in Theory and Research*. London: Routledge.

Bodé, Gérard. 1996. L'imposition de la langue (1789–1815). *Revue du Nord* 78, no. 317 (Oct.–Dec.): 771–80.

Bonder, Gloria. 1991. Research on Women in Latin America. In *Women's Studies International: Nairobi and Beyond*, edited by Aruno Rao, 135–41. New York: Feminist Press at the City University of New York.

Boneparth, Ellen. 1978. Evaluating Women's Studies. In *Women's Studies: An Interdisciplinary Collection*, edited by K. O. Blumhagen and W. D. Johnson, 21–30. Westport, Conn.: Greenwood.

Bonilla, Frank. 1991. *Brother Can We Paradigm?* New York: Inter-University Program on Latino Research.

Bonnell, Victoria E., and Lynn A. Hunt, eds. 1989. *The New Cultural History*. Berkeley and Los Angeles: University of California Press.

———, eds. 1999. *Beyond the Cultural Turn: New Directions in the Study of Society and Culture*. Berkeley and Los Angeles: University of California Press.

Bookchin, Murray. 1989. *Toward an Ecological Society*. Montreal: Black Rose Books.

———. 1991. *The Ecology of Freedom: The Emergence and Dissolution of Hierarchy*. New York: Black Rose Books.

Boring, Edwin Garrigues. [1929] 1950. *A History of Experimental Psychology*. 2nd ed. New York: Appleton-Century-Crofts.

———. 1961. *Psychologist at Large*. New York: Basic Books.

Bottazzini, Umberto. 1994. Geometry and "Metaphysics of Space" in Gauss and Riemann. In *Romanticism in Science*, edited by Stefano Poggi and Maurizio Bossi, 15–30. Dordrecht: Kluwer Academic.

Bowle, John. 1970. *The Unity of European History: A Political and Cultural Survey*. London: Oxford University Press.

Bowles, Gloria. 1983. Is Women's Studies an Academic Discipline? In *Theories of Women's Studies*, edited by Gloria Bowles and Renate Klein, 32–45. London: Routledge & Kegan Paul.

Bowles, Gloria, and Renate Klein, eds. 1983. *Theories of Women's Studies*. London: Routledge & Kegan Paul.

Boyer, Ernest L. 1987. *College: The Undergraduate Experience in America*. New York: Harper & Row.

Brain, Robert. 1991. The Geographical Vision and the Popular Order of Disciplines. In *World Views and Scientific Discipline Formation*, edited by William R. Woodward and Robert S. Cohen, 367–76. Dordrecht: Kluwer Academic.

Bramson, Leon. 1974. *The Political Context of Sociology*. Princeton, N.J.: Princeton University Press.

Bramwell, Anna. 1994. *The Fading of the Greens: The Decline of Environmental Politics in the West*. New Haven: Yale University Press.

Braudel, Fernand. 1984. *The Perspective of the World*. New York: Harper & Row.

Brockliss, L. W. B. 1992. The Scientific Revolution in France. In *The Scientific Revolution in National Context*, edited by Roy Porter and Mikulás Teich, 55–89. Cambridge: Cambridge University Press.

Bromwich, David. 1992. *Politics by Other Means: Higher Education and Group Thinking.* New Haven: Yale University Press.

Brown, Norman. 1971. Inaugural Session. In *Proceedings of the Twenty-seventh International Congress of Orientalists, Ann Arbor, 1967*, edited by Denis Sinor, 22–34. Wiesbaden: Otto Harrassowitz.

Buck, Peter. 1981. From Celestial Mechanics to Social Physics: Discontinuity in the Development of the Sciences in the Early Nineteenth Century. In *Epistemological and Social Problems of the Sciences in the Early Nineteenth Century*, edited by Hans Jahnke and Michael Otte, 19–33. Dordrecht: D. Reidel.

Buck, Philo M., Jr. 1930. *Literary Criticism: A Study of Values in Literature.* New York: Harper & Brothers.

Buhle, Mari Jo. 1998. *Feminism and Its Discontents: A Century of Struggle with Psychoanalysis.* Cambridge: Harvard University Press.

Bunch, Charlotte. 1986. Lesbians in Revolt. In *Women and Values: Readings in Recent Feminist Philosophy*, edited by M. Pearsall, 128–31. Belmont, Calif.: Wadsworth.

Burke, Peter. 1994. *Popular Culture in Early Modern Europe.* Aldershot, UK: Scolar Press.

Burrage, Michael. 1992. States as Users of Knowledge: A Comparison of Lawyers and Engineers in France and Britain. In *State Theory and State History*, edited by Rolf Torstendahl, 168–205. London: Sage.

Burwick, Frederick. 1986. *The Damnation of Newton: Goethe's Color Theory and Romantic Perception.* Berlin: Walter de Gruyter.

Bury, J. B. 1932. *The Idea of Progress: An Inquiry into Its Origins and Growth.* London: Macmillan.

Butler, Johnnella E. 1989. Difficult Dialogues. *Women's Review of Books* 6, no. 5:16.

Caban, Pedro. 1998. The New Synthesis of Latin American and Latino Studies. In *Borderless Borders: U.S. Latinos, Latin Americans, and the Paradox of Interdependence*, edited by Frank Bonilla, et al., 195–216. Philadelphia: Temple University Press.

Cain, William E., ed. 1994. *Teaching the Conflicts: Gerald Graff, Curricular Reform, and the Culture Wars.* New York: Garland.

Caldecott, Léonie, and Stephanie Leland, eds. 1983. *Reclaim the Earth: Women Speak Out for Life on Earth.* London: Women's Press.

Cardwell, Donald S. L. 1957. *The Organisation of Science in England: A Retrospect.* Melbourne: William Heinemann.

———. 1963. *Steam Power in the Eighteenth Century: A Case Study in the Application of Science.* London: Sheed & Ward.

Casti, John. 1994. *Complexification: Explaining a Paradoxical World through the Science of Surprise.* New York: HarperCollins.

Césaire, Aimé. [1955] 1972. *Discourse on Colonialism.* New York: Monthly Review.

Chabram, Angie. 1990. Chicana/o Studies as Oppositional Ethnography. *Cultural Studies* 4, no. 3 (Oct.): 228–47.

Chadwick, Owen. 1975. *The Secularization of the European Mind in the Nineteenth Century.* Cambridge: Cambridge University Press.

Chartier, Roger. 1989. Text, Printing, Readings. In *The New Cultural History: Essays*, edited by Lynn Hunt, 154–76. Berkeley and Los Angeles: University of California Press.

Chatterjee, Partha. 1993. *The Nation and Its Fragments: Colonial and Postcolonial Histories.* Princeton, N.J.: Princeton University Press.

Cheney, Lynne. 1988. *Humanities in America: A Report to the President, the Congress, and the American People.* Washington, D.C.: National Endowment for the Humanities.

———. 1990. *Tyrannical Machines: A Report on Education Practices Gone Wrong and Our Best Hopes for Setting Them Right*. Washington, D.C.: National Endowment for the Humanities.

Ching, Julia. 1978. Chinese Ethics and Kant. *Philosophy East and West* 28, no. 2:161–72.

Chodorow, Nancy. 1978. *The Reproduction of Mothering: Psychoanalysis and the Sociology of Gender*. Berkeley and Los Angeles: University of California Press.

Cixous, Hélène, and Catherine Clément. 1986. *The Newly Born Woman*. Minneapolis: University of Minnesota Press.

Clark, Stuart. 1983. French Historians and Early Modern Popular Culture. *Past and Present* 100:62–99.

Clarke, Desmond. 1989. *Occult Powers and Hypotheses: Cartesian Natural Philosophy under Louis XIV*. Oxford: Clarendon.

Clarke, J. J. 1997. *Oriental Enlightenment*. London: Routledge.

Cohen, Bernard I. 1985. *Revolution in Science*. Cambridge: Harvard University Press.

Cohen, Patricia Cline. 1984. Death and Taxes: The Domain of Numbers in Eighteenth-Century Popular Culture. In *Science and Technology in the Eighteenth Century: Essays of the Lawrence Gipson Institute for Eighteenth-Century Studies*, edited by Stephen H. Cutcliffe, 51–69. Bethlehem, Pa.: Lawrence Henry Gipson Institute.

Collard, Andree, and Joyce Contrucci. 1988. *Rape of the Wild*. London: Women's Press.

Collini Stefan, Donald Winch, and John Burrow. 1983. *The Noble Science of Politics: A Study in Nineteenth-Century Intellectual History*. Cambridge: Cambridge University Press.

Collins, H. M. 1983. An Empirical Relativist Programme in the Sociology of Scientific Knowledge. In *Science Observed: Perspectives on the Social Study of Science*, edited by Karin D. Knorr-Cetina and Michael Mulkay, 85–113. London: Sage.

Comte, Auguste. 1974. *The Positive Philosophy*. New York: AMS.

Connell, W. F. [1950] 1971. *The Educational Thought and Influence of Matthew Arnold*. Westport, Conn.: Greenwood.

Cook, Harold J. 1992. The New Philosophy in the Low Countries. In *The Scientific Revolution in National Context*, edited by Roy Porter and Mikuláš Teich, 115–49. Cambridge: Cambridge University Press.

Coronil, Fernando. 1996. Beyond Occidentalism: Towards Non-Imperial Geo-Historical Categories. *Cultural Anthropology* 1, no. 1:51–87.

———. 1997. *The Magical State: Nature, Money, and Modernity in Venezuela*. Chicago: University of Chicago Press.

Coulmas, Florian. 1995. Germanness: Language and Nation. In *The German Language and the Real World*, edited by Patrick Stevenson, 55–68. Oxford: Clarendon Press.

Court, Franklin. 1992. *Institutionalizing English Literature: The Culture and Politics of Literary Study, 1750–1900*. Stanford, Calif.: Stanford University Press.

Coutel, Charles. 1996. La Troisième République lit Condorcet. *Revue du Nord* 78, no. 317 (Oct.–Dec.): 967–74.

Cowan, George A., David Pines, and David Meltzer. 1994. *Complexity: Metaphors, Models, and Reality*. Reading, Mass.: Addison-Wesley.

Crawford, Elisabeth T. 1992. *Nationalism and Internationalism in Science, 1880–1939: Four Studies in the Nobel Population*. Cambridge: Cambridge University Press.

Crosby, Alfred W. 1997. *The Measure of Reality: Quantification and Western Society, 1250–1600*. Cambridge: Cambridge University Press.

Crosland, Maurice P. 1967. *The Society of Arcueil: A View of French Science at the Time of Napoleon I*. London: Heinemann Education Books.

———. 1975. The Development of a Professional Career in Science in France. *Minerva* 13, no. 1 (Spring): 38–57.

————. 1976. The Development of a Professional Career in Science in France. In *The Emergence of Science in Western Europe,* edited by Maurice P. Crosland, 139–59. New York: Science History Publications. (Also appeared in *Minerva* 13, no. 1 (Spring 1975): 38–57.)

Crosland, Maurice P., and Crosby Smith. 1978. The Transmission of Physics from France to Britain 1800–1840. *Historical Studies in the Physical Sciences* 9:1–61.

Crowther, J. G. 1966. *Statesmen of Science.* Chester Springs, Pa.: Dufour.

Culotta, Charles. 1974. German Biophysics, Objective Knowledge, and Romanticism. *Historical Studies in Physical Sciences* 4:3–38.

Cumings, Bruce. 1998. Boundary Displacement: Area Studies and International Studies during and after the Cold War. *Bulletin of Concerned Asian Scholars* 29, no. 1:6–26.

Cuomo, Chris J. 1998. *Feminism and Ecological Communities: An Ethic of Flourishing.* London: Routledge.

Curtis, Mark H., et al. 1985. *Integrity in the College Curriculum: A Report to the Academic Community.* Washington, D.C.: Association of American Colleges.

Dale, Peter Alan. 1989. *In Pursuit of a Scientific Culture: Science, Art, and Society in the Victorian Age.* Madison: University of Wisconsin Press.

Daly, Mary. 1984. *Pure Lust: Elemental Feminist Philosophy.* Boston: Beacon Press.

Dantzig, Tobias. 1968. *Henri Poincaré: Critic of Crisis: Reflections on His Universe of Discourse.* New York: Greenwood.

Darnton, Robert. 1984. *The Great Cat Massacre and Other Episodes in French Cultural History.* New York: Basic Books.

Daston, Lorraine. 1981. Mathematics and the Moral Sciences: The Rise and Fall of the Probability of Judgments, 1785–1840. In *Epistemological and Social Problems of the Sciences in the Early Nineteenth Century,* edited by Hans Jahnke and Michael Otte, 287–309. Dordrecht: D. Reidel.

————. 1988. *Classical Probability in the Enlightenment.* Princeton, N.J.: Princeton University Press.

————. 1992. The Doctrine of Chances without Chance: Determinism, Mathematical Probability, and Quantification in the Seventeenth Century. In *The Invention of Physical Science: Intersections of Mathematics, Theology, and Natural Philosophy since the Seventeenth Century,* edited by Mary Jo Nye, Joan L. Richards, and Roger H. Stewer, 27–50. Dordrecht: Kluwer Academic.

Dauben, Joseph. 1981. Mathematics in Germany and France in the Early Nineteenth Century: Transmission and Transformation. In *Epistemological and Social Problems of the Sciences in the Early Nineteenth Century,* edited by Hans Jahnke and Michael Otte, 371–99. Dordrecht: D. Reidel.

Davies, Mansel. 1947. *An Outline of the Development of Science.* London: Watts & Co.

Davies, Paul, ed. 1989. *The New Physics.* Cambridge: Cambridge University Press.

Davis, James. 1995. Philosophical Positivism and American Atonal Music Theory. *Journal of the History of Ideas* 56, no. 3 (July): 501–22.

Davis, Natalie Zemon. 1981. Anthropology and History in the 1980s: The Possibilities of the Past. *Journal of Interdisciplinary History* 12, no. 2:267–75.

————. 1983. *The Return of Martin Guerre.* Cambridge, Mass.: Harvard University Press.

Davis, Walter W. 1983. China, the Confucian Ideal, and the European Age of Enlightenment. *Journal of the History of Ideas* 44, no. 4:523–48.

Day, C. R. 1972–1973. Technical and Professional Education in France: The Rise and Fall of l'Enseignement Secondaire Spécial, 1865–1902. *Journal of Social History* 6, no. 2 (Winter): 177–201.

Dear, Peter R. 1995. *Discipline and Experience: The Mathematical Way in the Scientific Revolution.* Chicago: University of Chicago Press.

de Beauvoir, Simone. [1949] 1989. *The Second Sex*. New York: Vintage.

de Groot, J. J. M. 1912. *Religion in China: Universism—A Key to the Study of Taoism and Confucianism*. New York: Knickerbocker Press.

———. [1892] 1969. *The Religious System of China: Its Ancient Forms, Evolution, History and Present Aspect, Manners, Custom and Social Institution Connected Therewith*. Vol. 1. Taipei: Sing Man Press.

de Groot, Joanna, and Mary Maynard, eds. 1993. *Women's Studies in the 1990s: Doing Things Differently?* New York: St. Martin's.

Dejean, Joan. 1997. *Ancients against Moderns: Culture Wars and the Making of a Fin de Siècle*. Chicago: University of Chicago Press.

de la Campa, Román. 1999. *Latin Americanism*. Minneapolis: University of Minnesota Press.

Delesalle, Simone, and Jean-Claude Chevalier. 1986. *La linguistique, la grammaire et l'école 1750–1914*. Paris: Armand Colin.

Deleuze, Gilles. 1966. *Le Bergsonisme*. Paris: Presses Universitaires de France.

Delgado, Richard. 1995. *The Rodrigo Chronicles: Conversations about America and Race*. New York: New York University Press.

Demeritt, David. 1994. Ecology, Objectivity, and Critique in Writings on Nature and Human Societies. *Journal of Historical Geography* 20, no. 1:22–37.

den Boer, Pim. 1998. *History as a Profession*. Princeton, N.J.: Princeton University Press.

Descartes, René. 1968. *Discourse on Method and Other Writings*. Baltimore: Penguin.

———. 1980. *Discourse on Method and Meditations on First Philosophy*. Indianapolis: Hackett.

Devall, Bill, and George Sessions. 1985. *Deep Ecology: Living as if Nature Mattered*. Salt Lake City: Gibbs M. Smith.

de Zoysa, A., and C. D. Palitharatna. 1992. Models of European Scientific Expansion: A Comparative Description of "Classical" Medicine at the Time of Introduction of European Medicine to Sri Lanka, and Subsequent Development to the Present. In *Science and Empires: Historical Studies about Scientific Development and European Expansion*, edited by Patrick Petitjean, Catherine Jami, and Anne Marie Moulin, 111–20. Dordrecht: Kluwer Academic.

Dhombres, Nicole, and Jean Dhombres. 1989. *Naissance d'un nouveau pouvoir: Sciences et savants en France 1793–1824*. Paris: Editions Payot.

Diebolt, Claude. 1995. Le compte de l'éducation des universités en Prusse: 1868–1921. *Revue Historique* 294, no. 1 (July–Sept.): 85–107.

Dirlik, Arif. 1996. Chinese History and the Question of Orientalism. *History and Theory* 35, no. 4:96–118.

———. 1999. *After the Revolution: Waking to Global Capitalism*. Hanover, N.H.: Wesleyan University Press.

Dobson, Andrew, ed. 1991. *The Green Reader: Essay toward a Sustainable Society*. San Francisco: Mercury House.

———. 1992. *Green Political Thought: An Introduction*. London: Routledge.

———. 1995. *Green Political Thought*. New York: Routledge.

Dominiquez, Virginia R. 1989. *People as Subject, People as Object: Selfhood and Peoplehood in Contemporary Israel*. Madison: University of Wisconsin Press.

Drengson, Alan, and Yuichi Inoue, eds. 1995. *The Deep Ecology: An Introductory Anthology*. Berkeley, Calif.: North Atlantic Books.

Driver, F. 1992. Geography's Empire: Histories of Geographical Knowledge. *Environment and Planning D: Society and Space* 10, no. 1:23–40.

D'Souza, Dinesh. 1991. *Illiberal Education: The Politics of Race and Sex on Campus*. New York: Free Press.

DuBois, Ellen Carol, et al. 1985. *Feminist Scholarship: Kindling in the Groves of Academe.* Urbana: University of Illinois Press.

Du Bois, W. E. B. [1940] 1968. *Dusk of Dawn: An Essay toward an Autobiography of a Race Concept.* New York: Schocken Books.

Duchesneau, François. 1973. La philosophie Anglo-Saxonne de Bentham à W. James. In *La Philosophie du monde scientifique et industriel 1860 à 1940,* edited by François Chatelet, 123–50. Paris: Hachette.

Duggan, Lisa, and Nan D. Hunter. 1995. *Sex Wars: Sexual Dissent and Political Culture.* New York: Routledge.

Durant, Will. 1943. *The Story of Philosophy.* New York: Garden City Publishing.

During, Simon. 1993. Introduction. In *Cultural Studies: A Reader,* edited by Simon During, 1–25. New York: Routledge.

Durkheim, Emile. [1915] 1976. *The Elementary Forms of the Religious Life.* London: George Allen & Unwin.

Dussel, Enrique. 1995. *The Invention of the Americas: Eclipse of "the Other" and the Myth of Modernity.* New York: Continuum.

———. 1996. *The Underside of Modernity: Apel, Ricoeur, Rorty, and the Philosophy of Liberation.* New York: Humanities Press.

———. 1998. Beyond Eurocentrism: The World-System and the Limits of Modernity. In *The Cultures of Globalization,* edited by Frederic Jameson and Masao Miyoshi, 3–31. Durham, N.C.: Duke University Press.

Eagleton, Terry. 1983. *Literary Theory: An Introduction.* Oxford: Basil Blackwell.

Eban, Abba. 1961. Science and National Liberation. In *Science and the New Nations: The Proceedings of the International Conference on Science in the Advancement of New States at Rehovoth, Israel,,* edited by Ruth Gruber, 6–10. New York: Basic Books.

Eckersley, Robyn. 1992. *Environmentalism and Political Theory: Toward an Ecocentric Approach.* Albany: State University of New York Press.

Editors of *Lingua Franca,* eds. 2000. *The Sokal Hoax: The Sham That Shook the Academy.* Lincoln: University of Nebraska Press.

Edwards, Mark U. 1994. *Printing, Propaganda, and Martin Luther.* Berkeley and Los Angeles: University of California Press.

Egan, Rose F. 1921. The Genesis of the Theory of "Art for Art's Sake" in Germany and in England. *Smith College Studies in Modern Languages* 2, no. 4 (July): 5–61.

Ehrenreich, Barbara, and Deirdre English. 1973. *Witches, Midwives, and Nurses.* Old Westbury, N.Y.: Feminist Press.

Einstein, Albert. 1961. *Relativity: The Special and the General Theory.* New York: Crown.

———. 1968. Autobiographical Notes. In *Relativity Theory: Its Origins and Impact on Modern Thought,* edited by L. Pearce Williams, 85–93. New York: Wiley.

Eisenstein, Elizabeth L. 1983. *The Printing Revolution in Early Modern Europe.* Cambridge: Cambridge University Press.

Ekeland, Ivar. 1988. *Mathematics and the Unexpected.* Chicago: University of Chicago Press.

———. 1998. What Is Chaos Theory? *Review* 21, no. 2:137–50.

Engel, Arthur. 1983. The English Universities and Professional Education. In *The Transformation of Higher Learning 1860–1930,* edited by Konrad Jarausch, 293–305. Chicago: University of Chicago Press.

Engelhardt, Dietrich von. 1988. Romanticism in Germany. In *Romanticism in National Context,* edited by Roy Porter and Mikulás Teich, 109–33. Cambridge: Cambridge University Press.

Escobar, Arturo. 1995. *Encountering Development: The Making of the Third World.* Princeton, N.J.: Princeton University Press.

Evans, Judith. 1995. *Feminist Theory Today: An Introduction to Second-Wave Feminism.* London: Sage.

Evans, Mary. 1992. Feminism before Psychoanalysis. In *Feminism and Psychoanalysis: A Critical Dictionary,* edited by Elizabeth Wright, 98–103. Oxford: Blackwell.

———. 1997. Negotiating the Frontier: Women and Resistance in the Academy. In *Knowing Feminisms: On Academic Borders, Territories, and Tribes,* edited by Liz Stanley, 46–57. London: Sage.

Fairbank, John King. 1953. *Trade and Diplomacy on the China Coast: The Opening of Treaty Ports 1842–1854.* Cambridge: Harvard University Press.

———, et al. 1965. *East Asia: The Modern Transformation.* London: George Allen & Unwin.

Fancher, Raymond E. 1979. *Pioneers of Psychology: Studies of the Great Figures Who Paved the Way for the Contemporary Science of Behavior.* New York: W. W. Norton.

Fanon, Frantz. 1963. *Wretched of the Earth.* New York: Grove Weidenfeld.

———. 1967. *Black Skin, White Masks.* New York: Grove Weidenfeld.

Farmer, Doyne, et al., eds. 1986. *Evolution, Games, and Learning: Models for Adaptation in Machines and Nature, Proceedings of the Fifth Annual International Conference of the Center for Nonlinear Studies, Los Alamos, NM, May 20–24, 1985.* Special issue of the journal *Physica D: Nonlinear Phenomena 22,* nos. 1–3.

Farquhar, Judith B., and James L. Hevia. 1993. Culture and Postwar American Historiography of China. *Positions 1,* no. 2:486–525.

Farrar, W. V. 1976. Science and the German University System, 1790–1850. In *The Emergence of Science in Western Europe,* edited by Maurice P. Crosland, 179–92. New York: Science History Publications.

Fayet, Joseph. 1960. *La Révolution française et la science, 1789–1795.* Paris: Librairie Marcel Rivière.

Febvre, Lucien. [1941] 1953. Vivre l'histoire: Propos d'initiation. *Combats pour l'histoire,* 18–33. Paris: A. Colin.

Fechner, Gustav. [1860] 1966. *Elements of Psychophysics.* Vol. 1. New York: Holt, Rinehart & Winston.

Feigenbaum, Mitchell J. 1983. Universal Behavior in Nonlinear Systems. *Physica D: Nonlinear Phenomena 7,* no. 1–3: 16–39.

Fenton, William N. 1947. *Area Studies in American Universities: For the Commission on Implication of Armed Services Educational Programs.* Washington, D.C.: American Council on Education.

Fetterley, Judith. 1985. *Provisions: A Reader from Nineteenth-Century American Women.* Bloomington: Indiana University Press.

Feuer, Lewis. 1982. *Einstein and the Generations of Science.* New Brunswick, N.J.: Transactions.

Feyerabend, Paul. 1975. *Against Method.* London: New Left Books.

Figala, Karin, and Ulrich Petzgold. 1993. Alchemy in the Newtonian Circle: Personal Acquaintances and the Problem of the Late Phase of Isaac Newton's Alchemy. In *Renaissance and Revolution: Humanists, Scholars, Craftsmen and Natural Philosophers in Early Modern Europe,* edited by Judith Veronica Field and Frank A. J. L. James, 173–92. Cambridge: Cambridge University Press.

Filliozat, Jean. 1975. Project of Reform of the International Congresses of Orientalists. *Le XXIX Congrès International des Orientalistes, Paris, 1973,* edited by Committee of the Congress, 57–63. Paris: L'Asiathèque.

Flax, Jane. 1990. *Thinking Fragments: Psychoanalysis, Feminism, and Postmodernism in the Contemporary West.* Berkeley and Los Angeles: University of California Press.

Flexner, Eleanor. 1959. *A Century of Struggle: The Woman's Rights Movement in the United States.* Cambridge: Harvard University Press.

Fogel, Joshua A. 2000. *The Nanjing Massacre in History and Historiography*. Berkeley and Los Angeles: University of California Press.

Ford, Joseph. 1989. What Is Chaos and Should We Be Mindful of It? In *The New Physics*, edited by Paul Davies, 348–72. Cambridge: Cambridge University Press.

Forman, Paul. 1971. Weimar Culture, Causality, and Quantum Theory, 1918–1927: Adaptation by German Physicists and Mathematicians to a Hostile Intellectual Environment. *Historical Studies in the Physical Sciences* 3:1–115.

Foucault, Michel. 1977. *Language, Countermemory, Practice: Selected Essays and Interviews*. Translated by Donald F. Bouchard and Sherry Simon. Ithaca, N.Y.: Cornell University Press.

Fox, Robert. 1984. Science, the University, and the State in Nineteenth-Century France. In *Professions and the French State 1700–1900*, edited by Gerard Geison, 66–146. Philadelphia: University of Pennsylvania Press.

Fox, Stephen. 1985. *The American Conservation Movement: John Muir and His Legacy*. Madison: University of Wisconsin Press.

Franklin, Sarah, Celia Lury, and Jackie Stacey. 1991. Feminism and Cultural Studies: Pasts, Presents, Futures. In *Off-Centre: Feminism and Cultural Studies*, edited by Celia Lury and Jackie Stacey, 1–19. London: HarperCollins Academic.

Freedman, Maurice. 1979. On the Sociological Studies of Chinese Religions. In *The Study of Chinese Society: Essays by Maurice Freedman*, 351–72. Stanford, Calif.: Stanford University Press.

French, Marilyn. 1985. *Beyond Power: On Women, Men, and Morals*. New York: Summit Books.

French, Roger, and Andrew Cunningham. 1996. *Before Science: The Invention of the Friars' Natural Philosophy*. Aldershot, UK: Scolar.

Freud, Sigmund. 1957. *The Standard Edition of the Complete Psychological Works of Sigmund Freud*. Vol. 14 (1914–1916). London: Hogarth.

———. 1959. *The Standard Edition of the Complete Psychological Works of Sigmund Freud*. Vol. 20 (1925–1926). London: Hogarth.

Friedan, Betty. 1963. *The Feminine Mystique*. New York: Penguin.

Fuller, Steve. 1999. The Science Wars: Who Exactly Is the Enemy? *Social Epistemology* 13, no. 3/4:243–49.

———. 2000. Science Studies through the Looking Glass: An Intellectual Itinerary. In *Beyond the Science Wars: The Missing Discourse about Science and Society*, edited by Ullica Segerstråle, 185–217. Albany: State University of New York Press.

Furet, François, and Jacques Ozouf. 1982. *Reading and Writing: Literacy in France from Calvin to Jules Ferry*. Cambridge: Cambridge University Press.

Garcia, Mario T. 1989. *Mexican-Americans: Leadership, Ideology, Identity, 1930–1960*. New Haven: Yale University Press.

Garcia-Canclini, Nestor. 1990. *Culturas hibridas: Estrategias para entrar y salir de la modernidad*. Mexico: Grijalbo.

Gascoigne, John. 1989. *Cambridge in the Age of the Enlightenment: Science, Religion, and Politics from the Restoration to the French Revolution*. Cambridge: Cambridge University Press.

Gascoigne, Robert M. 1992. The Historical Demography of the Scientific Community, 1450–1900. *Social Studies of Science* 22:545–73.

Gascoigne, Robert Mortimer. 1987. *A Chronology of the History of Science, 1450–1900*. New York: Garland Publishing.

Gasking, Elizabeth. 1970. *The Rise of Experimental Biology*. New York: Random House.

Gates, Henry Louis. 1992. *Loose Canons: Notes on the Culture Wars*. New York: Oxford University Press.

Gaukroger, Steven, ed. 1980. *Descartes: Philosophy, Mathematics, and Physics*. Totowa, N.J.: Barnes & Noble.

Gaunt, Philip. 1996. Birmingham Revisited: Cultural Studies in a Post-Soviet World. *Journal of Popular Culture* 30, no. 1 (Summer): 91–102.

Gay, Peter. 1964. *The Enlightenment: An Interpretation*. Vol. 1, *The Rise of Modern Paganism*. New York: Norton.

———. 1969. *The Enlightenment: An Interpretation*. Vol. 2, *The Science of Freedom*. New York: Norton.

Geiger, Roger. 1980. Prelude to Reform: The Faculties of Letters in the 1860s. In *The Making of Frenchmen: History of Education in France, 1679–1979*, edited by Donald Baker and Patrick Harrigan, 337–61. Waterloo, Ont.: Historical Reflections Press.

Geiger, Roger L. 1986. *To Advance Knowledge: The Growth of American Research Universities 1900–1940*. Oxford: Oxford University Press.

Gell-Mann, Murray. 1988. The Concept of the Institute. In *The Emerging Syntheses in Science*, edited by David Pines, 1–15. Redwood City, Calif.: Addison-Wesley.

———. 1994. *The Quark and the Jaguar: Adventures in the Simple and the Complex*. New York: W. H. Freeman.

———. 1995. What Is Complexity? *Complexity* 1, no. 1:16–19.

Gemelli, Giuliana. 1999a. Enciclopedie ed enciclopedisti d'oltre Oceano tra l'età di Hoover e la guerra fredda. *Enciclopedie e scienze sociali nel XX secolo*, 135–86. Milano: Franco Angeli.

———. 1999b. Umanesimo e tolleranza: Progetti enciclopedici ed interscienza nell'Europa tra le due guerre. *Enciclopedie e scienze sociali nel XX secolo*, 23–47. Milano: Franco Angeli.

Gerbod, Paul. 1992. Aux origines de l'éducation technique en France (1750–1850). In *Chronological Section 1*. Seventeenth International Congress of Historical Sciences, Comité International des Sciences Historiques, Madrid, edited by Eloy Ruano and Manuel Burgos, 290–300.

Ghai, Dharam, and Jessica M. Vivian. 1992. *Grassroots Environmental Action: People's Participation in Sustainable Development*. London: Routledge.

Gieryn, Tom. 1996. Policing STS: A Boundary-Work Souvenir from the Smithsonian Exhibition on "Science in American Life." *Science, Technology, and Human Values* 21, no. 1 (Winter): 100–115.

Gilbert, Nigel G., and Michael Mulkay. 1984. *Opening Pandora's Box: A Sociological Analysis of Scientists' Discourse*. Cambridge: Cambridge University Press.

Gildea, Robert. 1980. Education and the Classes Moyennes in the Nineteenth Century. In *The Making of Frenchmen: History of Education in France, 1679–1979*, edited by Donald Baker and Patrick Harrigan, 275–99. Waterloo, Ont.: Historical Reflections Press.

Gilroy, Paul. 1987. *There Ain't No Black in the Union Jack*. London: Hutchinson.

———. 1993. *The Black Atlantic: Modernity and Double Consciousness*. Cambridge: Harvard University Press.

Ginzberg, Ruth. 1989. Uncovering Gynocentric Science. In *Feminism and Science*, edited by Nancy Tuana, 69–84. Bloomington: Indiana University Press.

Ginzburg, Carlo. 1980. *The Cheese and the Worms: The Cosmos of a Sixteenth-Century Miller*. New York: Dorset Press.

Giorgi, Amedeo. 1992. Toward the Articulation of Psychology as a Coherent Discipline. *A Century of Psychology as Science*, 46–59. Washington, D.C.: American Psychological Association.

Girardot, N. J. 1992. The Course of Sinological Discourse: James Legge (1815–97) and the Nineteenth-Century Invention of Taoism. In *Contacts between Cultures*, edited by Bernard Hung-Kay Luk and Barry D. Steben, 188–93. New York: Edwin Mellen Press.

Giroux, Henry A. 1992. Liberal Arts Education and the Struggle for Public Life: Dreaming about Democracy. In *Politics of Liberal Education,* edited by Darryl J. Gless and Barbara Herrnstein Smith, 119–44. Durham, N.C.: Duke University Press.

Gleick, James. 1987. *Chaos: Making a New Science.* New York: Viking.

Goldberg, David Theo. 1993. *Racist Culture: Philosophy and the Politics of Meaning.* Oxford: Blackwell.

————. 1994. Introduction: Multicultural Conditions. In *Multiculturalism. A Critical Reader,* 1–41. Cambridge, Mass.: Blackwell.

Goldman, Emma. N.d. *The Tragedy of Woman's Emancipation.* New York: Mother Earth Publishing.

Golinski, Jan. 1992. *Science as Public Culture: Chemistry and Enlightenment in Britain, 1760–1820.* Cambridge: Cambridge University Press.

Goodwin, C. James. 1985. On the Origins of Titchener's Experimentalists. *Journal of the History of Behavioral Sciences* 21, no. 4:383–89.

Gough, J. B. 1988. Lavoisier and the Fulfilment of the Stahlian Revolution. In *The Chemical Revolution: Essays in Reinterpretation,* edited by Arthur Donovan, 15–33. Philadelphia: University of Pennsylvania Press.

Gould, Stephen Jay. 1989. *Wonderful Life: The Burgess Shale and the Nature of History.* New York: W. W. Norton.

Gould, Stephen Jay, Norman L. Gilinsky, and Rebecca Z. German. 1987. Asymmetry of Lineages and the Direction of Evolutionary Time. *Science* 236, no. 4807 (12 June): 1437–41.

Gower, Barry. 1973. Speculation in Physics: The History and Practice of Naturphilosophie. *Studies in History and Philosophy of Science* 3, no. 4:301–56.

Grabiner, Judith V. 1981. Changing Attitudes toward Mathematical Rigor: Lagrange and Analysis in the Eighteenth and Nineteenth Centuries. In *Epistemological and Social Problems of the Sciences in the Early Nineteenth Century,* edited by Hans Jahnke and Michael Otte, 311–30. Dordrecht: D. Reidel.

Graff, Gerald. 1995. A Paradox of the Culture War. In *PC Wars: Politics and Theory in the Academy,* edited by Jeffrey Williams, 308–12. New York: Routledge.

Gray, Elizabeth Dodson. 1981. *Green Paradise Lost.* Wellesley, Mass.: Roundtable Press.

Gray, Ronald D. 1952. *Goethe the Alchemist: A Study of Alchemical Symbolism in Goethe's Literary and Scientific Works.* Cambridge: Cambridge University Press.

Greene, John C. 1981. *Science, Ideology, and World View: Essays in the History of Evolutionary Ideas.* Berkeley and Los Angeles: University of California Press.

Gregory, Frederick. 1977. *Scientific Materialism in Nineteenth-Century Germany.* Studies in the History of Modern Science, 1. Dordrecht: D. Reidel.

Grendler, Paul F. 1989. *Schooling in Renaissance Italy: Literacy and Learning, 1300–1600.* Baltimore: Johns Hopkins University Press.

Grier, P. T. 1990. Modern Ethical Theory and Newtonian Science. In *Philosophical Perspectives on Newtonian Science,* edited by Phillip Bricker and R. I. G. Hughes, 227–39. Cambridge: MIT Press.

Griffin, Susan. 1978. *Women and Nature: The Roaring Inside Her.* New York: Harper & Row.

Gripsrud, Jostein. 1989. "High Culture" Revisited. *Cultural Studies* 3:194–207.

Grosfoguel, Ramon. 1995. Global Logics in the Caribbean City System: The Case of Miami. In *World Cities in a World-System,* edited by Paul Knox and Peter J. Taylor, 156–70. Cambridge: Cambridge University Press.

Gross, Paul R. 1996. Reply to Tom Gieryn. *Science, Technology, and Human Values* 21, no. 1 (Winter): 116–20.

Gross, Paul R., and Norman Levitt. 1994. *Higher Superstition*. Baltimore: Johns Hopkins University Press.

Gross, Paul R., Norman Levitt, and Martin W. Lewis, eds. 1996. *The Flight from Science and Reason*. New York: New York Academy of Sciences.

Grossberg, Lawrence. 1993. The Formations of Cultural Studies: An American in Birmingham. In *Relocating Cultural Studies: Developments in Theory and Research*, edited by Valda Blundell, John Shepherd, and Ian Taylor, 21–66. London & New York: Routledge.

———. 1997. Cultural Studies, Modern Logics, and Theories of Globalisation. In *Back to Reality? Social Experience and Cultural Studies. Developments in Theory and Research*, edited by Angela McRobbie, 7–35. London: Routledge.

Grossberg, Lawrence, Cary Nelson, and Paula A. Treichler, eds. 1992. *Cultural Studies*. New York: Routledge.

Grosz, E. A., and Marie M. de Lepervanche. 1988. Feminism and Science. In *Crossing Boundaries: Feminisms and the Critique of Knowledges*, edited by Barbara Caine, E. A. Grosz, and Marie M. de Lepervanche, 5–27. Sydney: Allen & Unwin.

Guha, Ranajit. 1988a. On Some Aspects of the Historiography of Colonial India. In *Selected Subaltern Studies*, edited by Ranajit Guha and Gayatri Spivak, 37–44. New York: Oxford University Press.

———. 1988b. The Prose of Counter-Insurgency. In *Selected Subaltern Studies*, edited by Ranajit Guha and Gayatri Spivak, 45–86. New York: Oxford University Press.

Gulbenkian Commission on the Restructuring of the Social Sciences. 1996. *Open the Social Sciences: Report of the Gulbenkian Commission on the Restructuring of the Social Sciences*. Stanford, Calif.: Stanford University Press.

Guy, Basil. 1956. Rousseau and China. *Revue de Littérature Comparée* 30:531–36.

———. 1963. *The French Image of China before and after Voltaire*. Geneva: Institut et Musée Voltaire.

Guy-Sheftall, Beverly. 1991. A Black Feminist Perspective on the Academy. In *Transforming the Curriculum: Ethnic Studies and Women's Studies*, edited by Johnella E. Butler and John C. Walter, 305–12. Albany: State University of New York Press.

Hacking, Ian. 1983. Nineteenth-Century Cracks in the Concept of Determinism. *Journal of the History of Ideas* 44, no. 3 (July–Sept.): 455–75.

Hahn, Roger. 1967. Laplace as a Newtonian Scientist. Paper delivered at the Seminar on the Newtonian Influence held at Clark Library, University of California, Los Angeles, 8 April.

———. 1971. *The Anatomy of a Scientific Institution: The Paris Academy of Sciences, 1666–1803*. Berkeley and Los Angeles: University of California Press.

Haines, George. 1969. *Essays on German Influence upon English Education and Science, 1850–1919*. New London, Conn.: Archon Books and Connecticut College.

Hall, G. Stanley. 1923. *Life and Confessions of a Psychologist*. New York: D. Appleton.

Hall, Stuart. 1958. A Sense of Classlessness. *Universities and Left Review* 1, no. 5:26–32.

———. 1980. Cultural Studies: Two Paradigms. *Media, Culture, and Society* 2, no. 1 (Jan.): 57–72.

———. 1988. *The Hard Road to Renewal: Thatcherism and the Crisis of the Left*. London: Verso.

———. 1990. The Emergence of Cultural Studies and the Crisis of the Humanities. *October* 53:11–23.

———. 1992. Cultural Studies and Its Theoretical Legacies. In *Cultural Studies*, edited by Lawrence Grossberg, Cary Nelson, and Paula A. Treichler, 277–86. New York: Routledge.

————. 1996. What Is This "Black" Popular Culture? in *Stuart Hall: Critical Dialogues in Cultural Studies*, edited by David Morley and Kuan-Hsing Chen, 465–75. New York: Routledge.

Hall, Stuart, and Tony Jefferson, eds. 1976. *Resistance through Rituals*. London: Hutchinson.

Hall, Stuart, et al., eds. 1978. *Policing the Crisis*. London: Macmillan.

Hallam, Julia, and Annecka Marshall. 1993. Layers of Difference: The Significance of a Self-Reflexive Research Practice for a Feminist Epistemological Project. In *Making Connections: Women's Studies, Women's Movements, Women's Lives*, edited by Mary Kennedy, Cathy Lubelska, and Val Walsh, 64–78. London: Taylor & Francis.

Halls, W. D. 1965. *Society, Schools, and Progress in France*. Oxford: Pergamon.

Hamburger, Michael. 1957. *Reason and Energy: Studies in German Literature*. New York: Grove.

Hampson, Norman. 1990. *The Enlightenment*. East Rutherford, N.J.: Penguin.

Hanchard, Michael. 1990. Identity, Meaning, and the African-American. *Social Text 24* 8, no. 2:31–42.

Hanson, Norwood R. 1958. *Patterns of Discovery: An Inquiry into the Conceptual Foundations of Science*. Cambridge: Cambridge University Press.

Haraway, Donna Jeanne. 1981. In the Beginning Was the Word: The Genesis of Biological Theory. *Signs* 6, no. 3:469–81.

————. 1989. *Primate Visions: Gender, Race, and Nature in the World of Modern Science*. New York: Routledge.

————. 1991. *Simians, Cyborgs, and Women: The Reinvention of Nature*. London: Free Association.

Harding, Sandra G. 1986. *The Science Question in Feminism*. Ithaca, N.Y.: Cornell University Press.

————. 1987. *Feminism and Methodology: Social Science Issue*. Bloomington: Indiana University Press.

————. 1991. *Whose Science? Whose Knowledge? Thinking from Women's Lives*. Ithaca, N.Y.: Cornell University Press.

————. 1992. Why Physics Is a Bad Model for Physics. In *The End of Science: Attack and Defense*, edited by Richard Q. Elvee, 1–21. Lanham, Md.: University Press of America.

Harris, Jose. 1994. *Private Lives, Public Spirit: Britain 1870–1914*. Middlesex, UK: Penguin.

Harrison, Carol. 1999. *The Bourgeois Citizen in Nineteenth-Century France*. New York: Oxford University Press.

Hart, Gillian. 1991. Engendering Everyday Resistance: Gender, Patronage, and Production in Politics in Rural Malaysia. *Journal of Peasant Studies* 19, no. 1:93–121.

Haskell, Thomas L. 1977. *The Emergence of Professional Social Science: The American Social Science Association and the Nineteenth-Century Crisis of Authority*. Urbana: University of Illinois Press.

Hawking, Stephen W. 1988. *A Brief History of Time: From the Big Bang to Black Holes*. New York: Bantam.

Hebdige, Dick. 1979. *Subculture: The Meaning of Style*. London: Methuen.

Heilbron, Johan. 1990. *Het ontstaan van de sociologie*. Amsterdam: Prometheus. Also in English translation as *The Rise of Social Theory*, translated by Sheila Gogol. Minneapolis: University of Minnesota Press, 1995.

————. 1991. The Tripartite Division of French Social Science: A Long-Term Perspective. In *Discourses on Society: The Shaping of the Social Science Disciplines*, edited by

Peter Wagner, Björn Wittrock, and Richard Whitley, 73–92. Dordrecht: Kluwer Academic.

Herbst, Jurgen. [1965] 1972. *The German Historical School in American Scholarship: A Study in the Transfer of Culture*. Port Washington, N.Y.: Kennikat Press.

———. 1983. Diversification in American Higher Education. In *The Transformation of Higher Learning 1860–1930: Expansion, Diversification, Social Opening, and Professionalization in England, Germany, Russia, and the United States*, edited by Konrad H. Jarausch, 196–206. Chicago: University of Chicago Press.

Herf, Jeffrey. 1995. How the Culture Wars Matter: Liberal Historiography, German History, and the Jewish Catastrophe. In *Higher Education under Fire: Politics, Economics, and the Crisis of the Humanities*, edited by Michael Bérubé and Cary Nelson, 149–62. New York: Routledge.

Hertz, Frederic. 1962. *The Development of the German Public Mind: A Social History of German Political Sentiments, Aspirations, and Ideas*. New York: Macmillan.

Hicks, George. 1994. *The Comfort Women: Japan's Brutal Regime of Enforced Prostitution in the Second World War*. New York: W. W. Norton.

Hill, Christopher. 1964. *Society and Puritanism in Pre-Revolutionary England*. London: Secker & Warburg.

Hirsch, E. D. 1988. *Cultural Literacy: What Every American Needs to Know*. New York: Vintage.

Hobsbawm, Eric. 1959. *Primitive Rebels*. Manchester, UK: Manchester University Press.

———. 1980. The Revival of Narrative: Some Comments. *Past and Present*, no. 86:3–8.

———. 1989. *The Age of Empire: 1875–1914*. New York: Vintage.

Hoggart, Richard. 1957. *The Uses of Literacy*. London: Chatto & Windus.

Hohendahl, Peter Uwe. 1989. *Building a National Literature: The Case of Germany 1830–1870*. Ithaca, N.Y.: Cornell University Press.

Holton, Gerald. 1993. *Science and Anti-Science*. Cambridge: Harvard University Press.

Hood, Elizabeth F. 1984. Black Women, White Women: Separate Paths to Liberation. In *Feminist Frameworks: Alternative Theoretical Accounts of the Relations Between Women and Men*, edited by A. M. Jaggar and P. S. Rothenberg, 189–201. New York: McGraw-Hill.

hooks, bell. 1982. *Ain't I A Woman: Black Women and Feminism*. London: Pluto Press.

Hopkins, Terence K., Immanuel Wallerstein, et al., eds. 1996. *The Age of Transition: Trajectory of the World-System, 1945–2025*. London: Zed Books.

Horkheimer, Max, and Theodore Adorno. [1976] 1988. *Dialectic of Enlightenment*. Translated by John Cumming. New York: Continuum.

Howe, Florence. 1991. Women's Studies in the United States: Growth and Institutionalization. In *Women's Studies International: Nairobi and Beyond*, edited by Aruna Rao, 103–21. New York: Feminist Press at the City University of New York.

Høyrup, Jens. 1994. *In Number, Measure, and Weight: Studies in Mathematics and Culture*. Albany: State University of New York Press.

Hubbard, Ruth. 1988. Some Thoughts about the Masculinity of the Natural Sciences. In *Feminist Thought and the Structure of Knowledge*, edited by Mary M. Gergen, 1–15. New York: University Press.

———. 1989. Science, Facts, and Feminism. In *Feminism and Science*, edited by Nancy Tuana, 119–31. Bloomington: Indiana University Press.

Hull, Gloria T., Patricia Bell Scott, and Barbara Smith. 1982. *All the Women Are White, All the Blacks Are Men, but Some of Us Are Brave: Black Women's Studies*. Old Westbury, N.Y.: Feminist Press.

Hung, Ho-fung. 2003. Orientalist Knowledge and Social Theories: China and the European Conceptions of East-West Differences from 1600–1900. *Sociological Theory* 21, no. 3:254–80.

Hunter, James Davison. 1991. *Culture Wars: The Struggle to Define America.* New York: Basic Books.

Idema, Wilt. 1991. Preface. In *When West Meets East: International Sinology and Sinologists,* edited by Wang Jia Fong and Laura Li, 8–9. Taipei: Sinorama.

Iggers, Georg G. 1959. Further Remarks about Early Uses of the Term "Social Science." *Journal of the History of Ideas,* no. 20:433–36.

———. 1983. *The German Conception of History: The National Tradition of Historical Thought from Herder to the Present.* Middletown, Conn.: Wesleyan University Press.

———. 1995. Historicism: The History and Meaning of the Term. *Journal of the History of Ideas* 56, no. 1 (Jan.): 129–52.

———. 1997. *Historiography in the Twentieth Century: From Scientific Objectivity to the Postmodern Challenge.* Hanover, N.H.: Wesleyan University Press.

Ihsanoglu, Ekmeleddin. 1992. Ottomans and European Science. In *Science and Empires: Historical Studies about Scientific Development and European Expansion,* edited by Patrick Petitjean, Catherine Jami, and Anne Marie Moulin, 37–48. Dordrecht: Kluwer Academic.

Inkeles, Alex. 1969. Making Men Modern: On the Causes and Consequences of Individual Change in Six Developing Countries. *American Journal of Sociology* 75, no. 2:210–11.

Irigaray, Luce. 1985. *The Sex Which Is Not One.* Ithaca, N.Y.: Cornell University Press.

Jacob, Margaret C. 1981. *The Radical Enlightenment: Pantheists, Freemasons, and Republicans.* London: Allen & Unwin.

Jaggar, Alison M., and Paula S. Rothenberg, eds. 1984. *Feminist Frameworks: Alternative Theoretical Accounts of the Relations between Women and Men.* New York: McGraw-Hill.

Jahnke, Hans, and Michael Otte. 1981. On "Science as a Language." In *Epistemological and Social Problems of the Sciences in the Early Nineteenth Century,* edited by Hans Jahnke and Michael Otte, 75–90. Dordrecht: D. Reidel.

James, Frank. 1989. Introduction. In *The Development of the Laboratory,* edited by Frank A. J. L. James, 1–8. Basingstoke, UK: Macmillan Press Scientific & Medical.

Jami, Catherine. 1992. Western Mathematics in China, Seventeenth Century and Nineteenth Century. In *Science and Empires: Historical Studies about Scientific Development and European Expansion,* edited by Patrick Petitjean, Catherine Jami, and Anne Marie Moulin, 79–88. Dordrecht: Kluwer Academic.

Jarausch, Konrad. 1983. Higher Education and Social Change: Some Comparative Perspectives. In *The Transformation of Higher Learning 1860–1930: Expansion, Diversification, Social Opening, and Professionalization in England, Germany, Russia, and the United States,* edited by Konrad H. Jarausch, 9–36. Chicago: University of Chicago Press.

Jay, Gregory S. 1997. *American Literature and the Culture Wars.* Ithaca, N.Y.: Cornell University Press.

Jensen, Lionel M. 1997. *Manufacturing Confucianism: Chinese Traditions and Universal Civilization.* Durham, N.C.: Duke University Press.

Johnson, Richard. 1979. Culture and the Historians. In *Working Class Culture: Studies in History and Theory,* edited by John Clarke, C. Critcher, and Richard Johnson, 41–74. London: Hutchinson.

———. 1986–1987. What Is Cultural Studies Anyway? *Social Text,* no. 16 (Winter): 38–80.

———. 1991. Frameworks of Culture and Power: Complexity and Politics in Cultural Studies. In *Cultural Studies: Crossing Boundaries,* edited by R. Salper, 17–61. Atlanta: Rodopi.

Johnson-Odim, Cheryl. 1991. Common Themes, Different Contexts. In *Third World Women and the Politics of Feminism*, edited by Chandra Talpade Mohanty, Ann Russo, and Lourdes Torres, 314–27. Bloomington: Indiana University Press.

Jones, Ernest. 1953. *The Life and Work of Sigmund Freud*. Vol. 1, 1856–1900. New York: Basic Books.

———. 1955. *The Life and Work of Sigmund Freud*. Vol. 2, 1901–1919. New York: Basic Books.

Joseph, G. M., Catherine LeGrand, and Ricardo Donato Salvatore, eds. 1998. *Close Encounters of Empire: Writing the Cultural History of U.S.–Latin American Relations*. Durham, N.C.: Duke University Press.

Jungnickel, Christa. 1979. Teaching and Research in the Physical Sciences and Mathematics in Saxony, 1820–1850. *Historical Studies in the Physical Sciences*, no. 10:3–47.

Kaberry, Phyllis Mary. 1952. *Women of the Grassfields*. London: H.M. Stationery Office.

Kant, Immanuel. [1784] 1970. An Answer to the Question "What Is Enlightenment." In *Kant's Political Writings*, edited by Hans Reiss, 54–60. Cambridge: Cambridge University Press.

Kaplan, Amy, and Donald Pease, eds. 1993. *Cultures of United States Imperialism*. Durham, N.C.: Duke University Press.

Kaplan, Gisela, and Leslie Rogers. 1991. Biology and Feminism: Introduction. In *A Reader in Feminist Knowledge*, edited by Sneja Marina Gunew, 233–36. London: Routledge.

Kapp, K. William. [1950] 1970. *The Social Costs of Private Enterprise*. New York: Schocken Books.

Karim, Wazir-jahan. 1991. Research on Women in Southeast Asia: Current and Future Directions. In *Women's Studies International: Nairobi and Beyond*, edited by Aruna Rao, 142–55. New York: Feminist Press at the City University of New York.

Karp, Ivan. 1997. Does Theory Travel? Area Studies and Cultural Studies. *Africa Today* 44, no. 3:281–96.

Kaye, Harvey J. 1984. *British Marxist Historians: An Introductory Analysis*. Oxford, UK: Polity Press.

Keller, Evelyn Fox. 1982. Feminism and Sciences. *Signs* 7, no. 3:589–602.

———. 1989. The Gender/Science System: Or Is Sex to Gender as Nature Is to Science? In *Feminism and Science*, edited by Nancy Tuana, 33–44. Bloomington: Indiana University Press.

———. 1993. Fractured Images of Science, Language, and Power: A Postmodern Optic or Just Bad Eyesight? In *Knowledges: Historical and Critical Studies in Disciplinarity*, edited by Ellen Messer-Davidow, David R. Shumway, and David J. Sylvan, 54–69. Charlottesville: University of Virginia Press.

Keller, Evelyn Fox, and Christine R. Grontkowski. 1996. The Mind's Eye. In *Feminism and Science*, edited by Evelyn Fox Keller and Helen E. Longino, 187–202. Oxford: Oxford University Press.

Kelly-Gadol, Joan. 1976. The Social Relation of the Sexes: Methodological Implications of Women's History. *Signs*, no. 1:809–23.

Keohane, Nannerl. 1980. *Philosophy and the State in France: The Renaissance to the Enlightenment*. Princeton, N.J.: Princeton University Press.

Kerr, Clark. 1994. *The Uses of the University*. Cambridge: Harvard University Press.

Kevles, Daniel. 1977. *The Physicists: The History of a Scientific Community in America*. New York: Knopf.

Kieft, Robert. 1994. Cultural Studies: Part 1. *Choice: Current Reviews for Academic Libraries* 31, nos. 11–12 (July–Aug.): 1683–95.

Kiernan, Colm. 1973. *The Enlightenment and Science in Eighteenth-Century France*. Banbury, UK: Voltaire Foundation.

Kiernan, V. G. 1969. *The Lords of Human Kind: European Attitudes to the Outside World in the Imperial Age*. London: Weidenfeld & Nicolson.

Kim, Keong-il. 1997. Genealogy of the Idiographic vs. the Nomothetic Disciplines: The Case of History and Sociology in the United States. *Review* 20, no. 3/4, (Summer/Fall): 421–64.

Kim, Young Kun. 1978. Hegel's Criticism of Chinese Philosophy. *Philosophy East and West* 28, no. 2:173–80.

Kimball, Roger. 1990. *Tenured Radicals: How Politics Has Corrupted Our Higher Education*. New York: Harper & Row.

Klancher, Jon. 1989. Romantic Criticism and the Meanings of the French Revolution. *Studies in Romanticism* 28, no. 3 (Fall): 463–91.

Klein, Julia Thompson. 1990. *Interdisciplinarity: History, Theory, and Practice*. Detroit: Wayne State University Press.

Klemke, E. D. 1988. Introduction: What Is Philosophy of Science? In *Introductory Readings in the Philosophy of Science*, edited by Robert Hollinger and A. David Kline, 1–11. Buffalo, N.Y.: Prometheus.

Knight, David. 1970. The Physical Sciences and the Romantic Movement. *History of Science* 9: 54–75.

———. 1976. German Science in the Romantic Period. In *The Emergence of Science in Western Europe*, edited by Maurice Crosland, 161–78. New York: Science History Publications.

———. 1986. *The Age of Science: The Scientific World-View in the Nineteenth Century*. New York: Basil Blackwell.

Knock, Thomas J. 1992. *To End All Wars: Woodrow Wilson and the Quest for a New World Order*. Princeton, N.J.: Princeton University Press.

Knorr-Cetina, Karin D. 1981. *The Manufacture of Knowledge*. Oxford, UK: Pergamon.

———. 1983. The Ethnographic Study of Scientific Work: Towards a Constructivist Interpretation of Science. In *Science Observed*, edited by Karin D. Knorr-Cetina and Michael Mulkay, 115–40. London: Sage.

Koch, Sigmund, ed. 1959. *Psychology: A Study of a Science*. Vol. 1, *Sensory, Perceptual, and Physiological Formulations*. New York: McGraw-Hill.

———. 1961. Psychological Science versus the Science-Humanism Antinomy: Intimations of a Significant Science of Man. *American Psychologist* 16:629–39.

———. 1973. Theory and Experiment in Psychology. *Social Science Research* 40:669–709.

———. 1992. The Nature and Limits of Psychological Knowledge: Lessons of Century qua "Science." In *A Century of Psychology as Science*, edited by Sigmund Koch and David E. Leary, 7–35. Washington, D.C.: American Psychological Association.

Koot, Gerard M. 1980. English Historical Economics and the Emergence of Economic History in England. *History of Political Economy* 12, no. 2 (Summer): 174–205.

———. 1987. *English Historical Economics, 1870–1926: The Rise of Economic History and Neomercantilism*. Cambridge: Cambridge University Press.

Korg, Jacob. 1985. Astronomical Imaginary in Victorian Poetry. In *Victorian Science and Victorian Values: Literary Perspectives*, edited by James G. Paradis and Thomas Postlewait, 137–58. New Brunswick, N.J.: Rutgers University Press.

Kremer-Marietti, Angèle. 1982. *Le positivisme*. Vendôme: Presses Universitaires de France.

Krieger, Leonard. 1975. Elements of Early Historicism: Experience, Theory, and History in Ranke. *History and Theory* 14, no. 4 (Dec.): 1–14.

Krips, Henry. 1987. *The Metaphysics of Quantum Theory.* New York: Oxford University Press.

Kuhn, Thomas S. 1957. *Copernican Revolution.* Cambridge: Harvard University Press.

———. 1962. *The Structure of Scientific Revolutions.* Chicago: University of Chicago Press.

———. 1977. *The Essential Tension: Selected Studies in Scientific Tradition and Change.* Chicago: University of Chicago Press.

———. 1978. *Black-Body Theory and the Quantum Discontinuity, 1894–1912.* New York: Oxford University Press.

Kusch, Martin. 1995. *Psychologism: A Case Study in the Sociology of Philosophical Knowledge.* London: Routledge.

Lach, Donald F. 1977. *Asia in the Making of Europe.* Vol. 2, *A Century of Wonder.* Chicago: University of Chicago Press.

Laclau, Ernest, and Chantel Mouffe. 1985. *Hegemony and Socialist Strategy.* London: Verso.

Lakatos, Imre, and Alan Musgrave, eds. 1970. *Criticism and the Growth of Knowledge.* Cambridge: Cambridge University Press.

Lalouette, Jacqueline. 1998. La querelle de la foi et de la science et le banquet Berthelot. *Revue Historique* 300, no. 4 (Oct.–Dec.): 825–44.

Landes, Ruth. 1957. *The City of Women.* New York: Macmillan.

Larmore, Charles. 1980. Cartesian Optics and the Geometrization of Nature. In *Descartes: Philosophy, Mathematics, and Physics,* edited by S. Gaukroger, 6–22. Totowa, N.J.: Barnes & Noble.

Laszlo, Ervin. 1987. *Evolution: The Grand Synthesis.* Boston: Shambhala.

Latour, Bruno, and Steve Woolgar. 1979. *Laboratory Life: The Social Construction of Scientific Facts.* London: Sage.

———. [1979] 1986. *Laboratory Life: The Construction of Scientific Facts.* 2nd ed. Princeton, N.J.: Princeton University Press.

Lauter, Paul. 1995. "Political Correctness" and the Attack on American Colleges. In *Higher Education under Fire: Politics, Economics, and the Crisis of the Humanities,* edited by Michael Bérubé and Cary Nelson, 73–90. New York: Routledge.

LaVopa, Anthony. 1979. Status and Ideology: Rural Schoolteachers in pre-March and Revolutionary Prussia. *Journal of Social History* 12, no. 3 (Spring): 430–56.

Lawson, John, and Harold Silver. 1973. *A Social History of Education in England.* London: Methuen.

Le Roy Ladurie, Emmanuel. 1978. *Montaillou: The Promised Land of Error.* Translated by Barbara Bray. New York: George Braziller.

Leaf, Murray J. 1979. *Man, Mind, and Science: A History of Anthropology.* New York: Columbia University Press.

Lear, Linda J. 1993. Rachel Carson's Silent Spring. *Environmental History Review* 17, no. 2:23–48.

Leavis, F. R. 1930. *Mass Civilisation and Minority Culture.* Cambridge, UK: Minority Press.

Lee, Richard E. 1992. Readings in the "New Science": A Selective Annotated Bibliography. *Review* 15, no. 1 (Winter): 113–71.

———. 1994. Social Science Knowledge: A Report on Instituitionalization. Paper prepared for the Gulbenkian Commission on the Restructuring of the Social Sciences, Lisbon, June 17–18.

———. 1996. Structures of Knowledge. In *The Age of Transition: Trajectory of the World-System, 1945–2025,* edited by Terence K. Hopkins, Immanuel Wallerstein, et al., 179–206. London: Zed Books.

———. 2000. The Structures of Knowledge and the Future of the Social Sciences: Two Postulates, Two Propositions, and a Closing Remark. *Journal of World-Systems Research* 6, no. 3 (Fall/Winter): 786–96. http://jwsr.ucr.edu.

———. 2003. *Life and Times of Cultural Studies: The Politics and Transformation of the Structures of Knowledge.* Durham, N.C.: Duke University Press.

Lee, Richard E., and Immanuel Wallerstein. 2001. Structures of Knowledge. In *Blackwell Companion to Sociology,* edited by Judith R. Blau, 227–35. London: Blackwell.

Lees, Andrew. 1974. *Revolution and Reflection.* The Hague: M. Nijhoff.

Legge, Helen Edith. 1905. *James Legge: Missionary and Scholar.* London: Religious Tract Society.

Legge, James. [1880] 1976. *The Religions of China: Confucianism and Taoism Described and Compared with Christianity.* Folcroft, Pa.: Folcroft Library Editions.

———. [1891] 1957. *The Texts of Taoism.* New York: Julian Press.

Leites, Edmund. 1978. Confucianism in Eighteenth-Century England: Natural Morality and Social Reform. *Philosophy East and West* 28, no. 2:143–60.

Leith-Ross, Sylvia. 1939. *African Women: A Study of the Ibo of Nigeria.* New York: Faber.

Lele, Jayant. 1993. Orientalism and the Social Sciences. In *Orientalism and the Postcolonial Predicament: Perspective on South Asia,* edited by Carol A. Breckenridge and Peter van der Veer, 45–75. Philadelphia: University of Pennsylvania Press.

Lenoir, Timothy. 1993. The Discipline of Nature and the Nature of Disciplines. In *Knowledges: Historical and Critical Studies in Disciplinarity,* edited by Ellen Messer-Davidow, David R. Shumway, and David J. Sylvan, 70–102. Charlottesville: University of Virginia Press.

Lentini, Orlando. 1998. *La scienza sociale storica di Immanuel Wallerstein.* Milano: Franco Angeli.

Leopold, A. Starker. 1969. Editor's Foreword. In *The Subversive Sciences: Toward an Ecology of Man,* edited by Paul Shepard and Daniel McKinley, v–vi. New York: Houghton Mifflin.

Leopold, Aldo. [1949] 1969. Land Ethic. In *The Subversive Science: Essays Toward an Ecology of Man,* edited by Paul Shepard and Daniel McKinley, 402–15. New York: Houghton Mifflin.

Lepenies, Wolf. 1988. *Between Literature and Science: The Rise of Sociology.* Translated by R. J. Hollingdale. New York: Cambridge University Press.

Lerner, Gerda. 1972. *Black Women in White America: A Documentary History.* New York: Random House.

Lewis, John. 1974. *The Uniqueness of Man.* London: Lawrence & Wishart.

Li, Laura. 1991. Scholar of the Tao: French Sinologist Kristofer Schipper. In *When West Meets East: International Sinology and Sinologists,* edited by Wang Jia Fong and Laura Li, 112–21. Taipei: Sinorama.

Li, Tien-Yien, and James A. Yorke. 1975. Period Three Implies Chaos. *American Mathematical Monthly* 82, no. 10:985–92.

Lighthill, Sir James. 1986. The Recently Recognized Failure of Predictability in Newtonian Dynamics. *Proceedings of the Royal Society of London* 407, no. 1,832 (8 Sept.): 35–48.

Lindenberger, Herbert. 1989. On the Sacrality of Reading Lists: The Western Culture Debate at Stanford University. *Comparative Criticism,* 11:225–34.

Lindenfeld, David F. 1988. On Systems and Embodiments as Categories for Intellectual History. *History and Theory* 27:30–50.

———. 1993. The Myth of the Older Historical School of Economics. *Central European History* 26, no. 4:405–16.

———. 1997. *The Practical Imagination: The German Sciences of State in the Nineteenth Century.* Chicago: University of Chicago Press.

Linenthal, Edward T., and Tom Engelhardt. 1996. *History Wars: The Enola Gay and Other Battles for the American Past*. New York: Metropolitan Books.

List, Friedrich. [1841] 1856. *The National System of Political Economy*. Philadelphia: J. B. Lippincott.

Lloyd, Elisabeth Genevieve. 1996. Reason, Science, and the Domination of Matter. In *Feminism and Science*, edited by Evelyn Fox Keller and Helen E. Longino, 91–102. Oxford: Oxford University Press.

Loen, Arnold E. [1965] 1967. *Secularization: Science without God?* Philadelphia: Westminster Press.

Lorenz, Edward N. 1963a. Deterministic Nonperiodic Flow. *Journal of the Atmospheric Sciences* 20, no. 2 (Mar.): 130–41.

———. 1963b. The Mechanics of Vacillation. *Journal of the Atmospheric Sciences* 20, no. 5 (Sept.): 448–64.

———. 1964. The Problem of Deducing the Climate from the Governing Equations. *Tellus* 16, no. 1 (Feb.): 1–11.

Löwy, Michael, ed. 1982. *El Marxismo en America Latina (de 1909 a nuestros días)*. Mexico City: Era.

Lundgreen, Peter. 1983. Differentiation in German Higher Education. In *The Transformation of Higher Learning 1860–1930: Expansion, Diversification, Social Opening, and Professionalization in England, Germany, Russia, and the United States*, edited by Konrad H. Jarausch, 149–79. Chicago: University of Chicago Press.

Lutun, B. 1993. La survie d'une institution de l'Ancien Régime ou l'invention de l'École Polytechnique. *Revue Historique*, no. 586 (Apr.–June): 383–420.

Lux, David S. 1991. The Reorganization of Science 1450–1750. In *Patronage and Institutions: Science, Technology, and Medicine at the European Court 1500–1750*, edited by Bruce T. Moran, 185–94. Rochester, N.Y.: Boydell.

Lynch, Michael. 1985. *Art and Artifact in Laboratory Science*. London: Routledge.

Lynd, Robert S. 1940. *Knowledge for What?* New York: Harcourt, Brace.

Lyons, Henry. 1968. *The Royal Society, 1660–1940*. New York: Greenwood.

MacCabe, Colin. 1992. Cultural Studies and English. *Critical Quarterly* 34, no. 3:25–34.

Mach, Ernst. 1968. The Science of Mechanics. In *Relativity Theory: Its Origins and Impact on Modern Thought*, edited by L. Pearce Williams, 16–23. New York: Wiley.

Mackenzie, Brian. 1977. *Behaviorism and the Limits of Scientific Method*. London: Routledge & Kegan Paul.

MacLean, Michael. 1988. History in a Two-Cultures World: The Case of the German Historians. *Journal of the History of Ideas* 49, no. 3 (July–Sept.): 473–94.

MacLeod, Roy. 1981. Introduction: On the Advancement of Science. In *The Parliament of Science: The British Association for the Advancement of Science 1831–1981*, edited by Roy MacLeod and Peter Collins, 17–42. London: Science Reviews.

Makkreel, Rudolf, and Rodi Frithjof, eds. 1989. *Wilhelm Dilthey's Introduction to the Human Sciences*. Princeton, N.J.: Princeton University Press.

Mandelbrot, Benoit B. 1983. *The Fractal Geometry of Nature*. New York: W. H. Freeman.

Mandrou, Robert. 1979. *From Humanism to Science 1480–1700*. Atlantic Highlands, N.J.: Humanities Press.

Manicas, Peter T. 1987. *A History and Philosophy of the Social Sciences*. Oxford: Basil Blackwell.

———. 1991. The Social Science Disciplines: The American Model. In *Discourses on Society: The Shaping of the Social Science Disciplines*, edited by Peter Wagner, Björn Wittrock, and Richard Whitley, 45–71. Dordrecht: Kluwer Academic.

Manten, A. A. 1980. The Growth of European Scientific Journal Publishing before 1850. In *Development of Science Publishing in Europe*, edited by A. J. Meadows, 1–22. Amsterdam: Elsevier Science.

Mariátegui, José Carlos. [1928] 1971. *Seven Interpretive Essays on Peruvian Reality.* Austin: University of Texas.

Markov, Walter. 1989. *Napoleons keizerrijk: Geschiedenis en dagelijks leven na de Franse Revolutie.* Erfurt, Germany: De Walburg.

Markus, M., S. C. Müller, and G. Nicolis, eds. 1988. *From Chemical to Biological Organization.* Berlin: Springer-Verlag.

Marx, Karl. [1844] 1964. *The Economic and Philosophic Manuscripts of 1844,* edited by Dirk J. Struik. New York: International Publishers.

Mason, Stephen F. 1962. *A History of the Sciences.* New York: Collier.

Masterman, Margaret. 1970. The Nature of Paradigm. In *Criticism and the Growth of Knowledge,* edited by Imre Lakatos and Alan Musgrave, 59–90. Cambridge: Cambridge University Press.

Maurice, F. D. 1839. *Has the Church, or State, the Power to Educate the Nation?* London: Rivington & Darton & Clark.

McClellan, James E. 1985. *Science Reorganized: Scientific Societies in the Eighteenth Century.* New York: Columbia University Press.

McClelland, Charles. 1980. *State, Society, and University in Germany 1700–1914.* Cambridge: Cambridge University Press.

———. 1983. Professionalization and Higher Education in Germany. In *The Transformation of Higher Learning 1860–1930,* edited by Konrad Jarausch, 306–20. Chicago: University of Chicago Press.

McClelland, James. 1983. Diversification in Russian-Soviet Education. In *The Transformation of Higher Learning 1860–1930: Expansion, Diversification, Social Opening, and Professionalization in England, Germany, Russia, and the United States,* edited by Konrad H. Jarausch, 180–195. Chicago: University of Chicago Press.

McClintock, Anne. 1992. The Angel of Progress: Pitfalls of the Term "Post-Colonialism." *Social Text* 31/32:84–99.

McCormick, John. 1995. *The Global Environmental Movement.* 2nd ed. New York: John Wiley.

Mead, Margaret. 1935. *Sex and Temperament in Three Primitive Societies.* New York: Dell.

———. 1953. *Male and Female.* New York: Morrow.

Medvedev, Zhores. 1978. *Soviet Science.* New York: Norton.

Meeth, L. Richard. 1978. Interdisciplinary Studies: A Matter of Definition. *Change* 7: 10.

Menger, Carl. 1883. *Untersuchungen über die Methode der Sozialwissenschaften, und der Politischen Oekonomie insbesondere.* Berlin: Duncker & Humblot. Published as *Problems of Econoics and Sociology,* trans. F. J. Nock, ed. L. Schneider (Urbana: University of Illinois Press), 1963. Reprinted as *Investigations into the Method of the Social Sciences with Special Reference to Economics* (New York: New York University Press), 1985.

Merchant, Carolyn. 1980. *The Death of Nature.* San Francisco: HarperCollins.

———. 1989. *Ecological Revolution: Nature, Gender, and Science in New England.* Chapel Hill: University of North Carolina Press.

———. 1992. *Radical Ecology: The Search for a Livable World.* New York: Routledge.

Merrill, Lynn L. 1989. *The Romance of Victorian Natural History.* Oxford: Oxford University Press.

Merton, Robert K. 1979. *The Sociology of Science: An Episodic Memoir.* Carbondale: Southern Illinois University Press.

Meyerson, Emile. [1908] 1930. *Identity and Reality.* Translated by Kate Lowenbert. London: Allen & Unwin.

Mies, Maria, and Vandana Shiva. 1993. *Ecofeminism.* London: Zed Books.

Mignolo, Walter D. 1995. *The Darker Side of the Renaissance: Literacy, Territoriality, and Colonization.* Ann Arbor: University of Michigan Press.

———. 2000. *Local Histories/Global Designs: Coloniality, Subaltern Knowledges, and Border Thinking.* Princeton, N.J.: Princeton University Press.

Miller, Char, and Alaric V. Sample. 1998. Gifford Pinchot and the Conservation Spirit. In *Breaking New Ground,* edited by Gifford Pinchot, xi–xvii. Washington, D.C.: Island Press.

Miller, George A. 1992. The Constitutive Problem of Psychology. In *A Century of Psychology as Science,* edited by Sigmund Koch and David E. Leary, 40–45. Washington, D.C.: American Psychological Association.

Millett, Kate. 1971. *Sexual Politics.* London: Rupert Hart-Davis.

Mills, Charles K. 1997. *The Racial Contract.* Ithaca, N.Y.: Cornell University Press.

Mills, John A. 1998. *Control: A History of Behavioral Psychology.* New York: New York University Press.

Mirowski, Philip. 1989. *More Light than Heat: Economics as Social Physics, Physics as Nature's Economics.* Cambridge: Cambridge University Press.

Mitchell, Juliet. 1974. *Psychoanalysis and Feminism.* New York: Pantheon Books.

Mitzman, Arthur. 1996. Michelet and Social Romanticism: Religion, Revolution, Nature. *Journal of the History of Ideas* 57, no. 4 (Oct.): 659–82.

Moeller, Bernd, ed. 1972. *Imperial Cities and the Reformation.* Translated by H. C. Erik Midelfort and Mark U. Edwards Jr. Philadelphia: Fortress.

Mohanty, Chandra Talpade, Ann Russo, and Lourdes Torres, eds. 1991. *Third World Women and the Politics of Feminism.* Bloomington: Indiana University Press.

Moody, Joseph. 1978. *French Education since Napoleon.* Syracuse, N.Y.: Syracuse University Press.

Moraga, Cherrie, and Gloria Anzaldúa, eds. 1983. *This Bridge Called My Back: Writings by Radical Women of Color.* New York: Kitchen Table.

Moran, Bruce T. 1991. Patronage and Institutions: Courts, Universities, and Academies in Germany—An Overview, 1550–1750. In *Patronage and Institutions: Science, Technology, and Medicine at the European Court, 1500–1750,* edited by Bruce T. Moran, 169–85. Rochester N.Y.: Boydell.

Morange, Michel. 1998. *A History of Molecular Biology.* Translated by Matthew Cobb. Cambridge: Harvard University Press.

Moreiras, Alberto. 2001. *The Exhaustion of Difference: The Politics of Latin American Cultural Studies.* Durham, N.C.: Duke University Press.

Morgan, David. 1996. The Enchantment of Art: Abstraction and Empathy from German Romanticism to Expressionism. *Journal of the History of Ideas* 57, no. 2 (Apr.): 317–32.

Morrell, Jack, and Arnold Thackray. 1981. *Gentlemen of Science.* Oxford: Clarendon Press.

Morris, Meaghan. 1992. Reply to Hall. In *Cultural Studies,* edited by L. Grossberg, C. Nelson, and P. Treichler, 291–92. New York: Routledge.

Morus, Iwan. 1992. Different Experimental Lives: Michael Faraday and William Sturgeon. *History of Science* 30, part 1, no. 87 (Mar.): 1–28.

Moulines, Carlos-Ulises. 1981. Hermann von Helmholtz: A Physiological Approach to the Theory of Knowledge. In *Epistemological and Social Problems of the Sciences in the Early Nineteenth Century,* edited by Hans Jahnke and Michael Otte, 65–74. Dordrecht: D. Reidel.

Muchembled, Robert. 1985. *Popular Culture and Elite Culture in France 1400–1750.* Baton Rouge: Lousiana State University Press.

Müller, Friedrich Max. 1893. Inaugural Address. In *Transactions of the Ninth International Congress of Orientalists,* edited by E. Delmar Morgan, 1–37. London: Committee of the Congress.

Mungello, David E. 1985. *Curious Land: Jesuit Accommodation and the Origins of Sinology*. Stuttgart: Franz Steiner Verlag.

Muñoz, Carlos. 1989. *Youth, Identity, and Power: The Chicano Movement*. New York: Verso.

Musil, Caryn M., and Ruby Sales. 1991. Funding Women's Studies. In *Transforming the Curriculum: Ethnic Studies and Women's Studies*, edited by Johanna E. Butler and John C. Walter, 21–34. Albany: State University of New York Press.

Nakayama, Shigeru. 1982. Paradaimu-ron no Nijuu-nen (Two Decades of Paradigm Theory). *Kagaku* (Dec.): 770–71.

Nelson, Cary, and Lawrence Grossberg, eds. 1988. *Marxism and the Interpretation of Cultures*. London: Macmillan.

Nicolis, Grégoire, and Ilya Prigogine. 1989. *Exploring Complexity: An Introduction*. New York: W. H. Freeman.

Noble, David F. 1977. *America by Design: Science, Technology, and the Rise of Corporate Capitalism*. New York: Knopf.

Notz, William. 1926. Frederick List in America. *American Economic Review* 16:249–65.

Nye, Mary Jo. 1992. Physics and Chemistry: Commensurate or Incommensurate Sciences? In *The Invention of Physical Science: Intersections of Mathematics, Theology, and Natural Philosophy since the Seventeenth Century: Essays in Honor of Erwin N. Hierbert*, edited by Mary Jo Nye, Joan L. Richards, and Roger H. Steuwer, 205–24. Dordrecht: Kluwer Academic.

———. 1996. *Before Big Science: The Pursuit of Modern Chemistry and Physics, 1800–1940*. New York: Twayne.

O'Donnell, John M. 1985. *The Origins of Behaviorism: American Psychology, 1870–1920*. New York: New York University Press.

Ogburn William F., and Alexander Goldenweisser, eds. 1927. *The Social Sciences and Their Interrelations*. Boston: Houghton Mifflin.

Ohmann, Richard. 1991. Thoughts on Cultural Studies in the United States. In *Cultural Studies: Crossing Boundaries*, edited by Roberta Salper, special issue, *Critical Studies* 3, no. 1:5–15.

Olin, John C. 1994. *Erasmus, Utopia, and the Jesuits: Essays on the Outreach of Humanism*. New York: Fordham University Press.

Olson, Richard. 1990. *Science Deified and Science Defied: The Historical Significance of Science in Western Culture*. Vol. 2. Berkeley and Los Angeles: University of California Press.

Ortega y Gasset, José. 1968. The Historical Significance of the Theory of Einstein. In *Relativity Theory: Its Origins and Impact on Modern Thought*, edited by L. Pearce Williams, 147–57. New York: Wiley.

Ortiz, Fernando. [1940] 1963. *Contrapunteo cubano del tabaco y del azúcar*. Havana: Consejo Nacional de Cultura.

Outhwaite, William. 1996. The Philosophy of Social Science. In *The Blackwell Companion to Social Theory*, edited by Bryan S. Turner, 83–106. Oxford: Blackwell.

Ozouf, Mona. 1985. Unité nationale et unité de la pensée de Jules Ferry. In *Jules Ferry fondateur de la République*, edited by François Furet, 59–72. Paris: Éditions de l'École des Hautes Études en Sciences Sociales.

Pagels, Heinz R. 1988. *The Dreams of Reason: The Computer and the Rise of the Sciences of Complexity*. New York: Simon & Schuster.

Palat, Ravi. 1996. Fragmented Visions: Excavating the Future of Area Studies in a Post-American World. *Review* 29, no. 3:269–315.

Palmer, D. J. 1965. *The Rise of English Studies*. New York: Oxford University Press.

Pankhurst, Helen. 1992. *Gender, Development, and Identity*. London: Zed Books.

Parrinder, Patrick. 1991. *Authors and Authority*. London: Macmillan.

Patterson, Tiffany Ruby, and Robin D. G. Kelley. 2000. Unfinished Migrations: Reflections on the African Diaspora and the Making of the Modern World. *African Studies Review* 43, no. 1 (Apr.): 47–68.

Paul, Harry. 1985. *From Knowledge to Power.* New York: Cambridge University Press.

Paul, W. H. 1980. The Role and Reception of the Monograph in Nineteenth-Century French Science. In *Development of Science Publishing in Europe,* edited by A. J. Meadows, 123–48. Amsterdam: Elsevier Science.

Pavlova, Galina Evgen'evna, and S. R. Mikulinskii. 1990. *Organizatsiia nauki v Rossii v pervoi polovine XIX v.* (The Organization of Science in Russia in the First Half of the 19th Century.) Moscow: Nauka.

Peliti, L., and A. Vulpiani, eds. 1988. *Measures of Complexity: Proceedings of the Conference Held in Rome, September 30–October 2, 1987.* Berlin: Springer-Verlag.

Penley, Constance, and Andrew Ross, eds. 1991. *Technoculture.* Minneapolis: University of Minnesota Press.

Perez, Emma. 1999. *The Decolonial Imaginary: Writing Chicanas into History.* Bloomington: Indiana University Press.

Perez-Ramos, Antonio. 1988. *Francis Bacon's Idea of Science and the Makers' Knowledge Tradition.* Oxford: Clarendon.

Piasecki, Bruce. 1984. Sampler of Courses and Programs in Environmental Studies: Introduction to Environmental Education as Humanistic Inquiry. *Environmental Review* 8, no. 4 (Winter): 310–11.

———. 1992. Introduction to Special Issue: Environmental Coursework—Why We Have Assembled This Collection. *Environmental History Review* 16, no. 1 (Spring): 1–4.

Piscitelli, Alejandro. 1992. Neo-barroco, ciberpunks y la nueva frontera electronica. In *Tecnología y Modernidad en Latinoamerica: Etica, Politica y Cultura,* edited by Eduardo Sabrovsky, 1–7. Santiago, Chile: Ilet-Corfo, Hachette.

———. 1992–1993. Tecnologías de representación y epistemología de lo virtual. Buenos Aires: Ciclo Instituto Nacional de Antropología y Pensamiento Latinoamericano/Post-Grado en Antropología.

Pletsch, Carl. 1981. The Three Worlds, or the Division of Social Science Labor, circa 1950–1975. *Comparative Studies in Society and History* 23, no. 4:565–90.

Plotnitsky, Arkady. 1997. "But It Is Above All Not True": Derrida, Relativity, and the "Science Wars." *Postmodern Culture* 7, no. 2: www.iath.virginia.edu/pmc/contents.att.html.

Poggi, Stefano. 1994. Introduction. In *Romanticism in Science,* edited by Stefano Poggi and Maurizio Bossi, xi–xv. Dordrecht: Kluwer Academic.

Polanco, Xavier. 1992. World-Science: How Is the History of World-Science to Be Written? In *Science and Empires: Historical Studies about Scientific Development and European Expansion,* edited by Patrick Petitjean, Catherine Jami, and Anne Marie Moulin, 225–42. Dordrecht: Kluwer Academic.

Polanyi, Michael. 1958. *Personal Knowledge.* London: Routledge & Kegan Paul.

———. 1967. *The Tacit Dimension.* New York: Anchor.

Ponteil, Félix. 1966. *Histoire de l'enseignement en France, 1789–1965.* Paris: Sirey.

Poovey, Mary. 1998. *A History of the Modern Fact: Problems of Knowledge in the Science of Wealth and Society.* Chicago: University of Chicago Press.

Popper, Karl. 1959. *The Logic of Scientific Discovery.* London: Hutchinson.

———. 1961. *The Poverty of Historicism.* New York: Harper.

Porritt, Jonathon. 1985. *Seeing Green: The Politics of Ecology Explained.* New York: Basil Blackwell.

Prakash, Gyan. 1990. Writing Post-Orientalist Histories of the Third World: Perspectives from Indian Historiography. *Comparative Studies in Society and History* 32, no. 2:383–408.

————. 1992. Postcolonial Criticism and Indian Historiography. *Social Text*, no. 31/ 32:8–19.

Pratt, Mary Louise. 1992a. Humanities for the Future: Reflections on the Western Culture Debate at Stanford. In *The Politics of Liberal Education*, edited by Darryl J. Gless and Barbara Herrnstein Smith, 13–31. Durham, N.C.: Duke University Press.

————. 1992b. *Imperial Eyes: Travel Writing and Transculturation*. London: Routledge.

Preyer, Robert O. 1985. The Romantic Tide Reaches Trinity. In *Victorian Science and Victorian Values: Literary Perspectives*, edited by James Paradis and Thomas Postlewait, 39–68. New Brunswick, N.J.: Rutgers University Press.

Pribram, Karl A. 1992. Mind and Brain, Psychology and Neuroscience, the Eternal Verities. In *A Century of Psychology of Science*, edited by Sigmund Koch and David E. Leary, 700–720. Washington, D.C.: American Psychological Association.

Prigogine, Ilya. 1988. The New Convergence of Science and Culture. *UNESCO Courier* (May): 9–13.

————. 1996. *The End of Certainty: Time, Chaos, and the New Laws of Nature*. New York: Free Press.

Prigogine, Ilya, and Dean J. Driebe. 1997. Time, Chaos, and the Laws of Nature. In *Nonlinear Dynamics Chaotic and Complex Systems*, edited by E. R. Infeld, R. Zelazny, and A. Galkowski, 206–23. New York: Cambridge University Press.

Prigogine, Ilya, and Isabelle Stengers. 1984. *Order Out of Chaos: Man's New Dialogue with Nature*. New York: Bantam Books.

Prost, Antoine. 1977. *Histoire de l'enseignement en France, 1800–1967*. Paris: Armand Colin.

Pulleyblank, Edwin G. 1995. European Studies on Chinese Phonology: The First Phase. In *Europe Studies China: Papers from the International Conference on the History of European Sinology*, edited by Wilson Ming and John Cayley, 339–67. London: Han-Shan Tang Books.

Purver, Margery. 1967. *The Royal Society: Concept and Creation*. Cambridge: MIT Press.

Pyenson, Lewis. 1985. *Cultural Imperialism and Exact Sciences: German Overseas Expansion 1900–1930*. New York: Peter Lang.

————. 1993. *Civilizing Mission: Exact Sciences and French Overseas Expansion 1830–1940*. Baltimore: Johns Hopkins University Press.

Quijano, Aníbal. 1991. Colonialidad y modernidad/racionalidad. *Peru Indigena* 29, no. 13.

————. 2000a. Colonialidad del poder y clasificación social. *Journal of World-Systems Research* 6, no. 2 (Summer/Fall): 342–86. http://jwsr.ucr.edu.

————. 2000b. Coloniality of Power, Ethnocentrism, and Latin America. *Nepantla* 1, no. 3:139–55.

Rabkin, Jeremy. 1999. The Culture War That Isn't. *Policy Review*, no. 96 (Aug.–Sept.): 3–19.

Radcliffe, Sarah, and Sallie Westwood. 1996. *Remaking the Nation: Place, Identity, and Politics in Latin America*. London: Routledge.

Ragland-Sullivan, E. 1982. Jacques Lacan: Feminism and the Problem of Gender Identity. *Sub-Stance*, no. 36:6–20.

————. 1992. Jacques Lacan. In *Feminism and Psychoanalysis: A Critical Dictionary*, edited by Elizabeth Wright, 201–7. Oxford: Blackwell.

Ramos, Julio. 1989. *Desencuentros de la modernidad en America Latina: Literatura y politica en el siglo XIX*. Mexico City: Fondo de Cultura Economica.

Rand, Ayn. 1969. *The Romantic Manifesto: A Philosophy of Literature*. Cleveland: World.

Randall, John H. 1976. *The Making of the Modern Mind*. New York: Columbia University Press.

Ranke, Leopold von. 1973. *The Theory and Practice of History*, edited by Georg G. Iggers and Konrad von Moltke. Indianapolis: Bobbs-Merrill.

Rao, Aruna, ed. 1991. *Women's Studies International: Nairobi and Beyond.* New York: Feminist Press at the City University of New York.

Readings, Bill. 1996. *The University in Ruins.* Cambridge: Harvard University Press.

Reichwein, Adolf. 1925. *China and Europe: Intellectual and Artistic Contacts in the Eighteenth Century.* New York: Barnes & Noble.

Reisner, Edward H. 1925. *Nationalism and Education since 1789.* New York: Macmillan.

Ribe, Neil. 1985. Goethe's Critique of Newton: A Reconsideration. *Studies in History and Philosophy of Science* 16, no. 4 (Dec.): 315–35.

Rice, Stuart A., ed. 1931. *Methods in Social Science: A Case Book.* Chicago: University of Chicago Press.

Richard, Nelly. 1989. *La estratificación de los márgenes.* Santiago, Chile: Malvern.

———. 1993. The Latin American Problematic of Theoretical-Cultural Transference: Postmodern Appropriations and Counterappropriations. *South Atlantic Quarterly* 92, no. 3 (Summer): 453–60.

———. 1994. *La insubordinación de los signos: Cambio politico, transformaciones culturales y poéticas de la crisis.* Santiago, Chile: Editorial Cuarto Propio.

Ride, Lindsay. 1960. Biographic Note. In *The Chinese Classics* by James Legge, 1025. Hong Kong: Hong Kong University Press.

Rietbergen, Peter. 1998. *Europe: A Cultural History.* London: Routledge.

Ringer, Fritz. 1967. Higher Education in Germany in the Nineteenth Century. *Journal of Contemporary History* 2, no. 3 (July): 123–38.

Roberts, Gerrylynn K. 1991. Scientific Academies across Europe. In *The Rise of Scientific Europe,* edited by David Goodman and Colin Russell, 227–52. Sevenoaks, UK: Hodder & Stoughton.

Rocheleau, Dianne, Barbara Thomas-Slayter, and Esther Wangari, eds. 1996. *Feminist Political Ecology: Global Issues and Local Experiences.* London: Routledge.

Rocke, Alan. 1993. *The Quiet Revolution: Hermann Kolbe and the Science of Organic Chemistry.* Berkeley and Los Angeles: University of California Press.

Roderick, Gordon, and Michael Stephens. 1972. *Scientific and Technical Education in Nineteenth-Century England.* Newton Abbot, UK: David & Charles.

Rodo, José Enrique. 1988. *Ariel.* Austin: University of Texas Press.

Rollin, Bernard. 1998. *The Unheeded Cry.* Ames: Iowa State University Press.

Roos, D. A. 1985. The "Aims and Intentions" of Nature. In *Victorian Science and Victorian Values: Literary Perspectives,* edited by James Paradis and Thomas Postlewaite, 159–80. New Brunswick, N.J.: Rutgers University Press.

Rosaldo, Michelle Z., Louise Lamphere, and Joan Bamberger. 1974. *Women, Culture, and Society.* Stanford, Calif.: Stanford University Press.

Rosenzweig, Saul. 1992a. Freud and Experimental Psychology: The Emergence of Idiodynamics. In *A Century of Psychology as Science,* edited by Sigmund Koch and David E. Leary, 135–207. Washington, D.C.: American Psychological Association.

———. 1992b. *Freud, Jung, and Hall the King-Maker: The Historic Expedition to America (1909).* St. Louis, Mo.: Rana House.

Roshwald, Aviel, and Richard Stites, eds. 1999. *European Culture in the Great War.* New York: Cambridge University Press.

Ross, Andrew. 1996a. Introduction. *Social Text* 14, nos. 1 & 2 (Spring/Summer): 1–13.

———. 1996b. *Science Wars.* Durham, N.C.: Duke University Press.

Ross, Dorothy. 1979. The Development of the Social Sciences. In *The Organization of Knowledge in Modern America, 1860–1920,* edited by A. Oleson and J. Voss, 107–38. Baltimore: Johns Hopkins University Press.

———. 1991. *The Origins of American Social Science.* New York: Cambridge University Press.

Ross, Ralph. 1976. On the International Encyclopaedia of the Social Sciences. *American Political Science Review* 70, no. 3:939–51.

Rosser, Sue Vilhauer. 1992. *Biology and Feminism: A Dynamic Interaction.* New York: Twayne.

Rothblatt, Sheldon. 1983. The Diversification of Higher Education in England. In *The Transformation of Higher Learning 1860–1930,* edited by Konrad Jarausch, 131–48. Chicago: University of Chicago Press.

Rothman, Tony. 1997. Irreversible Differences. *Sciences* 37, no. 4 (July–Aug.): 26–31.

Roversi, Antonio. 1984. *Il magistero della scienza: Storia del Verein für Sozialpolitik dal 1872 al 1888.* Milan: Angeli.

Rowbotham, Arnold H. 1942. *Missionary and Mandarin: The Jesuits at the Court of China.* New York: Russell & Russell.

Rudy, Willis. 1984. *The Universities of Europe, 1100–1914.* London: Associated University Presses.

Ruelle, David, and F. Takens. 1971. On the Nature of Turbulence. *Communications in Mathematical Physics* 20:167–92.

Rüger, Adolf, and Helmut Klein, eds. 1985. *Humboldt-Universitat zu Berlin: Uberblick 1810–1985.* Berlin: VEB Deutscher Verlag der Wissenschaften.

Rule, Paul A. 1986. *K'ung-tzu or Confucius? The Jesuit Interpretation of Confucianism.* London: Allen & Unwin.

Rupke, N. A. 1983. The Study of Fossils in the Romantic Philosophy of History and Nature. *History in Science* 21, part 4, no. 54 (Dec.): 389–413.

Russell, Bertrand. 1954. *Mysticism and Logic and Other Essays.* Suffolk, UK: Penguin.

Russell III, Edmund P. 1997. Lost among the Parts per Billion: Ecological Protection at the United States Environmental Protection Agency, 1970–1993. *Environmental History* 2, no. 1 (Jan.): 29–51.

Saarinen, A. 1988. Feminist Research: In Search of a New Paradigm. *Acta Sociologica* 31, no. 1:35–51.

Sabean, David. 1984. *Power in the Blood: Popular Culture and Village Discourse in Early Modern Germany.* New York: Cambridge University Press.

Sachs, Mendel. 1988. *Einstein versus Bohr: The Continuing Controversies in Physics.* La Salle, Ill.: Open Court.

Sadker, Myra, and Nancy Frazier. 1973. *Sexism in School and Society.* New York: Harper & Row.

Said, Edward W. 1978. *Orientalism.* New York: Pantheon.

Saldaña, Juan-José. 1992. Science et pouvoir au XIXe siècle: La France et le Mexique en perspective. In *Science and Empires: Historical Studies about Scientific Development and European Expansion,* edited by Patrick Petitjean, Catherine Jami, and Anne Marie Moulin, 153–64. Dordrecht: Kluwer Academic.

Saldivar, José David. 1991. *The Dialectics of Our America: Genealogy, Cultural Critique, and Literary History.* Durham, N.C.: Duke University Press.

———. 1997. *Border Matters: Remapping American Cultural Studies.* Berkeley and Los Angeles: University of California Press.

Saldivar-Hull, Sonia. 2000. *Feminism on the Border: Chicana Gender Politics and Literature.* Berkeley and Los Angeles: University of California Press.

Salomon-Bayet, Claire. 1981. 1802—"Biologie" et Medicine. In *Epistemological and Social Problems of the Sciences in the Early Nineteenth Century,* edited by Hans Jahnke and Michael Otte, 35–54. Dordrecht: D. Reidel.

Salvatore, Ricardo D. 1998. The Enterprise of Knowledge: Representational Machines of Informal Empire. In *Close Encounters of Empire: Writing the Cultural History of*

U.S.–Latin American Relations, edited by Gilbert M. Joseph and R. D. Salvatore, 69–106. Durham, N.C.: Duke University Press.

Samson, Anne. 1992. *Leavis*. Toronto: University of Toronto Press.

Sandoval, Chela. 1991. U.S. Third World Feminism: The Theory and Method of Oppositional Consciousness in the Postmodern World. *Genders*, no. 10 (Spring): 1–24.

———. 2000. *Methodology of the Oppressed*. Minneapolis: University of Minnesota Press.

Santiago-Valles, Kelvin A. 1994. *"Subject People" and Colonial Discourses: Economic Transformation and Social Disorder in Puerto Rico 1898–1947*. Albany: State University of New York Press.

Sarlo, Beatriz. 1994. *Escenas de la vida posmoderna*. Madrid: Espasa Calpe.

Sarton, Georges. 1924. The New Humanism. *Isis* 6, no. 1:9–42.

Sasaki, Chikara. 1994. Kagakushi no Kouryuu (The Rise of the History of Science). *Iwanami Kouza Gendai Shisou* 10, Kagaku-ron (Iwanami Shoten): 271–313.

Sauer, Wolfgang, and Helmut Glück. 1995. Norms and Reforms: Fixing the Form of the Language. In *The German Language and the Real World*, edited by Patrick Stevenson, 69–93. Oxford: Clarendon Press.

Schipper, Kristofer. 1995. The History of Taoist Studies in Europe. In *Europe Studies China: Papers from the International Conference on the History of European Sinology*, edited by Wilson Ming and John Cayley, 467–91. London: Han-Shan Tang Books.

Schlesinger, Arthur M., Jr. 1992. *The Disuniting of America: Reflections on a Multicultural Society*. New York: W. W. Norton.

Schmitt, Charles B. 1984. *The Aristotelian Tradition and Renaissance Universities*. London: Variorum.

Schmoller, Gustav von. 1883. Zur Methodologie der Staats- und Sozialwissenschaften. *Jahrbuch für Gesetzgebung, Verwaltung und Volkswirschaft im Deutschen Reich*: 974–94. Leipzig: n.p.

Schramm, Sarah Slavin. 1978. Women's Studies: Its Focus, Idea, Power, and Promise. In *Women's Studies: An Interdisciplinary Collection*, edited by Kathleen O'Connor Blumhagen and Walter D. Johnson, 3–12. Westport, Conn.: Greenwood.

Schumacher, Ernst Friedrich. 1973. *Small Is Beautiful: Economics As If People Mattered*. New York: Harper & Row.

Schuster, Marilyn R., and Susan R. Van Dyne. 1985. *Women's Place in the Academy: Transforming the Liberal Arts Curriculum*. Totowa, N.J.: Rowman & Allanheld.

Schwab, Raymond. 1984. *The Oriental Renaissance: Europe's Rediscovery of India and the East 1680–1880*. New York: Columbia University Press.

Schweber, S. S. 1985. Scientists as Intellectuals: The Early Victorians. In *Victorian Science and Victorian Values: Literary Perspectives*, edited by James Paradis and Thomas Postlewait, 1–37. New Brunswick, N.J.: Rutgers University Press.

Scolum, Sally. 1975. Woman the Gatherer: Male Bias in Anthropology. In *Toward an Anthropology of Women*, edited by Rayna R. Reiter, 36–50. New York: Monthly Review.

Scott, Joan W. 1995. The Rhetoric of Crisis in Higher Education. In *Higher Education under Fire: Politics, Economics, and the Crisis of the Humanities*, edited by Michael Bérubé and Cary Nelson, 293–304. New York: Routledge.

Sears, Robert R. 1992. Psychoanalysis and Behavior Theory: 1907–1965. In *A Century of Psychology as Science*, edited by Sigmund Koch and David E. Leary, 208–20. Washington, D.C.: American Psychological Association.

Segerstråle, Ullica. 2000a. Science and Science Studies: Enemies or Allies? In *Beyond the Science Wars: The Missing Discourse about Science and Society*, edited by Ullica Segerstråle, 1–40. Albany: State University of New York Press.

———. 2000b. History of Science: Stirred, Not Shaken. *Science* 290, no. 5497:n.p. http://sciencemag.org.

Seligman, Edwin Robert Anderson. 1925. Economics in the United States: An Historical Sketch. In *Essays in Economics*, 122–60. New York: Macmillan.

Senghaas, Dieter. 1985. *The European Experience: A Historical Critique of Development Theory*. Dover, N.H.: Berg.

———. 1991. Friedrich List and the Basic Problems of Modern Development. *Review* 14, no. 3 (Summer): 451–67.

Senn, Peter R. 1958. The Earliest Use of the Term "Social Science." *Journal of the History of Ideas* 19, no. 4 (Oct.): 568–70.

Sewell, William H., Jr. 1985. Ideologies and Social Revolutions: Reflections on the French Case. *Journal of Modern History* 57, no. 1:57–85.

Shapin, Steven. 1982. History of Science and Its Sociological Reconstruction. *History of Science* 20:157–211.

———. 1994. *A Social History of Truth: Civility and Science in Seventeenth-Century England*. Chicago: University of Chicago Press.

———. 1995. Here and Everywhere: Sociology of Scientific Knowledge. *Annual Review of Sociology* 21:289–321.

———. 1996. *The Scientific Revolution*. Chicago: University of Chicago Press.

Shapin, Steven, and Simon Schaffer. 1985. *Leviathan and the Air-Pump: Hobbes, Boyle, and the Experimental Life*. Princeton, N.J.: Princeton University Press.

Shapiro, Fred R. 1984. A Note on the Origin of the Term "Social Science." *Journal of the History of the Behavioral Sciences* 20, no. 1 (Jan.): 20–22.

Shaw, Robert. 1981. Strange Attractors, Chaotic Behavior, and Information Flow. *Zeitschrift für Naturforschung* 36a, no. 1:80–112.

Shepard, Paul. 1969. Introduction: Ecology and Man—A Viewpoint. In *The Subversive Science: Essays toward an Ecology of Man*, edited by P. Shepard and D. McKinley, 1–10. New York: Houghton Mifflin.

Sheridan, Susan. 1991. Feminist Knowledge and Women's Studies: Introduction. In *A Reader in Feminist Knowledge*, edited by Sneja Marina Gunew, 45–47. London: Routledge.

Shinn, Terry. 1979. The French Science Faculty System, 1808–1914: Institutional Change and Research Potential in Mathematics and the Physical Sciences. *Historical Studies in the Physical Sciences* 10: 271–332.

Shiva, Vandana. 1989. *Staying Alive: Women, Ecology and Development*. London: Zed Books.

Shor, Ira. 1992. *Culture Wars: School and Society in the Conservative Restoration*. Chicago: University of Chicago Press.

Skinner, William. 1964. What the Study of China Can Do for Social Science. *Journal of Asian Studies* 23, no. 4:517–22.

Smith, Barbara Herrnstein. 1992. Cult-Lit: Hirsch, Literacy, and the "National Culture." In *The Politics of Liberal Education*, edited by Darryl J. Gless and Barbara Herrnstein Smith, 75–94. Durham, N.C.: Duke University Press.

Smith, Crosbie. 1998. *The Science of Energy: A Cultural History of Energy in Victorian Britain*. Chicago: University of Chicago Press.

Smith, Herbert F. 1965. *John Muir*. New York: Twayne.

Smith, Jonathan. 1994. *Fact and Feeling: Baconian Science and the Nineteenth-Century Literary Imagination*. Madison: University of Wisconsin Press.

Snelders, H. A. M. 1970. Romanticism and Naturphilosophie and the Inorganic Natural Sciences 1797–1840: An Introductory Survey. *Studies in Romanticism* 9, no. 3 (Summer): 191–215.

———. 1994. *Wetenschap en intuitie: Het Duits romantisch-speculatief natuuronderzoek rond 1800*. Baarn, Netherlands: Ambo.

Snell, Marilyn Berlin, and Yevgeny Yevtushenko. 1992. Beyond Being and Becoming. *New Perspectives Quarterly* 9, no. 2 (Spring): 22–28.

Snow, C. P. 1959. *Two Cultures and the Scientific Revolution.* Cambridge: Cambridge University Press.

———. 1993. *The Two Cultures.* Cambridge: Cambridge University Press.

Société Asiatique de Paris, ed. 1949. *Actes du XXI Congrès International des Orientalistes.* Paris: Imprimerie Nationale.

Sommers, Christina Hoff. 1994. *Who Stole Feminism? How Women Have Betrayed Women.* New York: Simon & Schuster.

Southern, R. W. 1953. *The Making of the Middle Ages.* New Haven: Yale University Press.

Sparks, Colin. 1996. Stuart Hall, Cultural Studies, and Marxism. In *Stuart Hall: Critical Dialogues in Cultural Studies,* edited by Stuart Hall, David Morley, and Kuan-Hsing Chen, 71–111. London: Routledge.

Spence, Jonathan D. 1992. Looking East: The Long View. In *Chinese Roundabout: Essays in History and Culture,* 78–92. New York: W. W. Norton.

———. 1998. *The Chan's Great Continent: China in Western Minds.* New York: W. W. Norton.

Spivak, Gayatri Chakravorty. 1988. Subaltern Studies: Deconstructing Historiography. In *Selected Subaltern Studies,* edited by Ranajit Guha and Gayatri Chakravorty Spivak, 3–35. New York & Oxford: Oxford University Press.

———. 1991. Reflections on Cultural Studies in the Post-colonial Conjuncture: An Interview with the Guest Editor. In *Cultural Studies: Crossing Boundaries,* edited by Roberta Salper, special issue, *Critical Studies* 3, no. 1:63–78.

Spretnak, Charlene, and Fritjof Capra. 1986. *Green Politics: The Global Promise.* Santa Fe, N.Mex.: Bear.

Stachel, John. 1998. *Einstein's Miraculous Year: Five Papers That Changed the Face of Physics.* Princeton, N.J.: Princeton University Press.

Stam, Robert, and Ella Shohat. 1994. *Unthinking Eurocentrism: Multiculturalism and the Media.* New York: Routledge.

Stanton, Domna C., and Abigail J. Stewart, eds. 1995. *Feminisms in the Academy.* Ann Arbor: University of Michigan Press.

Stavenhagen, Rodolfo. 1973. Siete tesis equivocadas sobre America Latina. In *Tres ensayos sobre America Latina,* edited by Rodolfo Stavenhagen, Ernesto Laclau, and Ruy Mauro Marini. Barcelona: Editorial Anagrama.

Stebbing, Susan L. 1944. *Philosophy and the Physicists.* Middlesex, UK: Penguin.

Stein, Daniel L., ed. 1989. *Lectures in the Sciences of Complexity: The Proceedings of the 1988 Complex Systems Summer School held June–July 1988 in Santa Fe, New Mexico.* Redwood City, Calif.: Addison-Wesley.

Stepan, Nancy Leys. 1996. Race and Gender: The Role of Analogy in Science. In *Feminism and Science,* edited by Evelyn Fox Keller and Helen E. Longino, 121–36. Oxford: Oxford University Press.

Stephenson, R. H. 1995. *Goethe's Conception of Knowledge and Science.* Edinburgh: Edinburgh University Press.

Stimpson, Catharine R. 1992. Afterword: Lesbian Studies in the 1990's. In *Lesbian Texts and Contexts: Radical Revisions,* edited by Karla Jay and Joanne Glasgow, 377–82. London: Onlywomen.

Stimpson, Catharine R., and Nina Kressner Cobb. 1986. *Women's Studies in the United States.* New York: Ford Foundation.

Stone, Lawrence. 1979. The Revival of Narrative: Reflections on a New Old History. *Past and Present,* no. 85:3–24.

Stone, Norman. 1999. *Europe Transformed, 1878–1919.* Oxford: Blackwell.

Stratton, Jon, and Ien Ang. 1996. On the Impossibility of a Global Cultural Studies: British Cultural Studies in an International Frame. In *Stuart Hall: Critical Dialogues in Cultural Studies,* edited by Stuart Hall, David Morley, and Kuan-Hsing Chen, 61–91. London: Routledge.

Striphas, Ted. 1998a. Introduction: The Long March—Cultural Studies and Its Institutionalization. *Cultural Studies* 12, no. 4:453–75.

———. 1998b. Cultural Studies Institutional Presence: A Resource and Guide. *Cultural Studies* 12, no. 4:571–94.

Strohmayer, Ulf. 1997. The Displaced, Deferred, or Was It Abandoned Middle: Another Look at the Idiographic-Nomothetic Distinction in the German Social Sciences. *Review* 20, no. 3/4:279–344.

Sussman, Héctor J., and Raphael S. Zahler. 1975. Catastrophe Theory as Applied to the Social and Biological Sciences: A Critique. *Synthese* 31, no. 2 (Aug.): 117–216.

Sutton, Geoffrey V. 1995. *Science for a Polite Society: Gender, Culture, and the Demonstration of Enlightenment.* Boulder, Colo.: Westview.

Sykes, Charles T. 1988. *ProfScam: Professors and the Demise of Higher Education.* Washington, D.C.: Regnery Gateway.

Talbott, John. 1969. *The Politics of Education Reform in France, 1918–1940.* Princeton, N.J.: Princeton University Press.

Teeter-Dobbs, Betty Jo, and Margaret Jacob. 1995. *Newton and the Culture of Newtonianism.* Atlantic Highlands, N.J.: Humanities.

Thom, René. 1975. *Structural Stability and Morphogenesis.* Reading, Mass.: Benjamin.

Thompson, E. P. 1966. *The Making of the English Working Class.* New York: Vintage.

———. 1978. *The Poverty of Theory and Other Essays.* London: Merlin Press.

———. 1991. *Customs in Common.* New York: New Press.

Thrift, Nigel. 1996. *Spatial Formations.* London: Sage.

Titchener, Edward Bradford. 1899. Structural and Functional Psychology. *Philosophical Review* 8:290–99.

Tolstoy, Ivan. 1990. *The Knowledge and the Power: Reflections on the History of Science.* Edinburgh: Canongate.

Tong, Rosemarie. 1989. *Feminist Thought: A Comprehensive Introduction.* Boulder, Colo.: Westview.

Townson, Michael. 1992. *Mother-Tongue and Fatherland: Language and Politics in German.* Manchester, UK: Manchester University Press.

Trachtman, Leon E., and Robert Perrucci. 2000. *Science under Siege? Interest Groups and the Science Wars.* Lanham, Md.: Rowman & Littlefield.

Trebilcot, Joyce. 1984. Sex Roles: The Argument from Nature. In *Feminist Frameworks: Alternative Theoretical Accounts of the Relations between Women and Men,* edited by Alison M. Jaggar and Paula S. Rothenberg. New York: McGraw-Hill.

Tribe, Keith. 1995. *Strategies of Economic Order: German Economic Discourse, 1750–1950.* Cambridge: Cambridge University Press.

Trompf, G. W. 1978. *Friedrich Max Mueller: As a Theorist of Comparative Religion.* Bombay: Shakuntala Publishing House.

Tsing, Lowenhaupt. 1993. *In the Realm of Diamond Queen.* Princeton, N.J.: Princeton University Press.

Turner, Bryan S. 1978. *Marx and the End of Orientalism.* London: Allen & Unwin.

Turner, Graeme. 1990. *British Cultural Studies: An Introduction.* Boston: Unwin Hyman.

Turner, Stephen Park, and Jonathan H. Turner. 1990. *The Impossible Science: An Institutional Analysis of American Sociology.* Newbury Park, Calif.: Sage.

Turner, Steven. 1971. The Growth of Professorial Research in Prussia, 1818–1848: Causes and Context. *Historical Studies in the Physical Sciences* 3:137–82.

———. 1981. The Prussian Professoriate and the Research Imperative, 1790–1840. In

Epistemological and Social Problems of the Sciences in the Early Nineteenth Century, edited by Hans Jahnke and Michael Otte, 109–21. Dordrecht: D. Reidel.

Van de Graaff, John. 1973. The Politics of German University Reform 1810–1970. PhD diss., Columbia University.

Van Tassel, David D. 1984. From Learned Society to Professional Organization: The American Historical Association, 1884–1900. *American Historical Review* 89, no. 4 (Oct.): 929–56.

Verdenal, R. 1973. A. A. Cournot. In *L'Histoire de la philosophie,* vol. 6, *La Philosophie du monde scientifique et industriel 1860 à 1940,* edited by François Châtelet. Paris: Hachette.

Verger, Jacques, ed. 1986. *Histoire des universités en France.* Toulouse: Bibliothèque historique Privat.

Vloemans, Antoon. 1966. *Bergson.* The Hague, Netherlands: Kruseman.

Voegelin, Eric. 1975. *From Enlightenment to Revolution,* edited by John Hallowell. Durham, N.C.: Duke University Press.

———. 1998. *History of Political Ideas, 4: Renaissance and Reformation,* edited by David A. Morse and William M. Thompson. Columbia: University of Missouri Press.

Waerden, B. L. van der, ed. 1967. *Sources of Quantum Mechanics.* New York: Dover.

Wagner, Peter. 1991. Science of Society Lost: On the Failure to Establish Sociology in Europe during the "Classical" Period. In *Discourses on Society: The Shaping of the Social Science Disciplines,* edited by Peter Wagner, Björn Wittrock, and Richard Whitley, 219–45. Dordrecht: Kluwer Academic.

Waldrop, M. Mitchell. 1992. *Complexity: The Emerging Science at the Edge of Order and Chaos.* New York: Touchstone/Simon & Schuster.

Wall, Derek. 1994. *Green History: A Reader in Environmental Literature, Philosophy, and Politics.* London: Routledge.

Wallerstein, Immanuel. 1974. The Rise and Future Demise of the World Capitalist System: Concepts for Comparative Analysis. *Comparative Studies in Society and History* 16, no. 4 (Sept.): 387–415.

———. 1977. The Tasks of Historical Social Science: An Editorial. *Review* 1, no. 1 (Summer): 3–7.

———. 1989a. *The Modern World-System.* Vol. 3, *The Second Era of Great Expansion of the Capitalist World-Economy, 1730's–1840's.* New York: Academic Press.

———. 1989b. The French Revolution as a World-Historical Event. *Social Research* 57, no. 1 (Spring): 33–52.

———. 1991a. *Geopolitics and Geoculture: Essays on the Changing World-System.* Cambridge: Cambridge University Press.

———. 1991b. *Unthinking Social Science: The Limits of Nineteenth-Century Paradigms.* Cambridge, UK: Polity Press.

———. 1993. Annales School. In *The Blackwell Dictionary of Twentieth-Century Social Thought,* edited by William Outhwaite and Tom Bottomore, 16–18. Oxford: Basil Blackwell.

———. 1996. History in Search of Science. *Review* 19, no. 1 (Winter): 11–22.

———. 1997a. The Unintended Consequences of Cold War Area Studies. In *The Cold War and the University: Toward an Intellectual History of the Postwar Years,* edited by Noam Chomsky et al., 195–232. New York: New Press.

———. 1997b. Social Science and the Quest for a Just Society. *American Journal of Sociology* 102, no. 5 (Mar.): 1241–57.

Warren, Karen, ed. 1996. *Ecological Feminist Philosophies.* Bloomington: Indiana University Press.

Watanabe, Masao. 1976. *The Japanese and Western Science.* Translated by O. T. Benfey. Philadelphia: University of Pennsylvania Press.

Weaver, Warren. 1948. Science and Complexity. *American Scientist* 36, no. 4:536–44.

Weber, Eugen. 1976. *Peasants into Frenchmen: The Modernization of Rural France, 1870–1914*. Stanford, Calif.: Stanford University Press.

Weber, Max. [1922] 1951. *The Religion of China: Confucianism and Taoism*. Glencoe, Ill.: Free Press.

Webster, Charles. 1982. *From Paracelsus to Newton: Magic and the Making of Modern Science*. Cambridge: Cambridge University Press.

Weinberg, Steven. 1992. *Dreams of a Final Theory*. New York: Pantheon.

———. 1994. Life in the Universe. *Scientific American* 271, no. 4 (Oct.): 44–49.

———. 1996. Sokal's Hoax. *New York Review of Books*, Aug. 8, 11–12; 14–15.

Weinberger, Jerry. 1985. *Science, Faith, and Politics: Francis Bacon and the Utopian Roots of the Modern Age*. Ithaca, N.Y.: Cornell University Press.

Weiner, Myron. 1966. *Modernization: The Dynamics of Growth*. Voice of America Forum Lectures. New York: Basic.

Weitz, Morris, ed. 1966. *Twentieth-Century Philosophy: The Analytic Tradition*. New York: Free Press.

West, Guida, and Rhoda Lois Blumberg. 1990. *Women and Social Protest*. Oxford: Oxford University Press.

Westfall, Richard S. 1977. *The Construction of Modern Science: Mechanisms and Mechanics*. Cambridge: Cambridge University Press.

White, Reginald. 1970. *The Anti-philosophers: A Study of the Philosophers in Eighteenth-Century France*. London: Macmillan.

Whitehead, Alfred N. 1947. *Essays in Science and Philosophy*. New York: Philosophical Library.

Widmalm, Sven. 1992. Instituting Science in Sweden. In *The Scientific Revolution in National Context*, edited by Roy Porter and Mikulás Teich, 240–62. Cambridge: Cambridge University Press.

Williams, Jeffrey, ed. 1995. *PC Wars: Politics and Theory in the Academy*. New York: Routledge.

Williams, L. Pearce. 1973. Kant, Naturphilosophie, and Scientific Method. In *Foundations of Scientific Method: The Nineteenth Century*, edited by Ronald N. Giere and Richard S. Westfall, 3–22. Bloomington: Indiana University Press.

Williams, L. Pearce, and Henry John Steffens. 1978. *The History of Science in Western Civilization*. Vol. 3, *Modern Science, 1700–1900*. Washington, D.C.: University Press of America.

Williams, Raymond. 1958. *Culture and Society*. London: Chatto & Windus.

———. 1977. *Marxism and Literature*. Oxford: Oxford University Press.

Wilshire, Bruce. 1989. *The Moral Collapse of the University: Professionalism, Purity, and Alienation*. Albany: State University of New York Press.

Wilson, Edward O. 1975. *Sociobiology: A New Synthesis*. Cambridge: Harvard University Press.

Wilson, Norman. 1999. *History in Crisis? Recent Directions in Historiography*. Upper Saddle River, N.J.: Prentice Hall.

Windschuttle, Keith. 1997. *The Killing of History: How Literary Critics and Social Theorists Are Murdering Our Past*. New York: Free Press.

Wolf, Eric. 1982. *Europe and the People without History*. Berkeley and Los Angeles: University of California Press.

Wolpert, Lewis. 1993. *The Unnatural Nature of Science*. Cambridge: Harvard University Press.

Women's Studies Group, CCCS. 1978. *Women Take Issue*. London: Hutchinson.

Wood, Ellen Meiksins, and John Bellamy Foster. 1997. *In Defense of History: Marxism and the Postmodern Agenda*. New York: Monthly Review.

Woodhouse, A. S. P. 1968. The New Role of the Humanities in Canada. In *Of Several Branches: Essays from the Humanities Association Bulletin*, edited by Gerald Mc-Caughey and Maurice Legris, 50–64. Toronto: University of Alberta by University of Toronto Press.

Woolf, Virginia. 1929. *A Room of One's Own*. New York: Harcourt & Brace.

———. 1938. *Three Guineas*. London: Hogarth.

Woster, Donald. 1991. *Nature's Economy*. Cambridge: Cambridge University Press.

Wright, Elizabeth, ed. 1992. *Feminism and Psychoanalysis: A Critical Dictionary*. Oxford: Blackwell.

Yates, Frances. 1964. *Giordano Bruno and the Hermetic Tradition*. London: Routledge & Kegan Paul.

Yeo, Richard. 1981. The Scientific Method and the Image of Science. In *The Parliament of Science: The British Association for the Advancement of Science 1831–1981*, edited by Roy MacLeod and Peter Collins, 65–88. London: Science Reviews.

Yoshiaki, Yoshimi. 2000. *Sexual Slavery in the Japanese Military during World War II*. New York: Columbia University Press.

Young, Robert. 1990. *White Mythologies: Writing History and the West*. London: Routledge.

———. 1996. *Two Halves: Political Conflict in Literary and Cultural Theory*. Manchester, UK: Manchester University Press.

Yúdice, George, Jean Franco, and Juan Flores, eds. 1992. *The Crisis of Contemporary Latin American Culture*. Minneapolis: University of Minnesota Press.

Zeeman, E. C. 1977. *Catastrophe Theory: Selected Papers, 1972–1977*. Reading, Mass.: Addison-Wesley.

Zhang, Long-Xi. 1988. The Myth of the Other: China in the Eyes of the West. *Critical Inquiry* 15, no. 1:108–31.

Zupko, Ronald. 1990. *Revolution in Measurement: Western European Weights and Measures since the Age of Science*. Philadelphia: American Philosophical Society.

Zurcher, Erik. 1994. Conception, Birth, and Early Childhood of the Documentation Center. *China Information: Anniversary Supplement* 9, no. 1:10–16.

Index

About the Contributors

Volkan Aytar is with the Turkish Economic and Social Studies Foundation (TESEV) in Istanbul.

Ayşe Betül Çelik is assistant professor of political science and with the Conflict Analysis and Resolution Program at Sabancı University, Turkey.

Mauro Di Meglio is assistant professor of sociology at the University of Naples, Italy.

Mark Frezzo is assistant professor of sociology at Florida Atlantic University.

Ho-fung Hung is assistant professor of sociology at the Chinese University of Hong Kong.

Biray Kolluoğlu Kırlı is assistant professor of sociology at Bogaziçi University, Turkey.

Agustín Lao-Montes is assistant professor of sociology at the University of Massachusetts–Amherst.

Richard E. Lee is deputy director of the Fernand Braudel Center and associate professor of sociology at Binghamton University. He is the author of *Life and Times of Cultural Studies: The Politics and Transformation of the Structures of Knowledge*.

Eric Mielants is assistant professor of sociology at the University of Utah.

Boris Stremlin is a doctoral candidate in sociology at Binghamton University.

Sunaryo is a doctoral candidate in sociology at Binghamton University.

Immanuel Wallerstein is director of the Fernand Braudel Center and senior research scholar at Yale University. He is the author of *The Modern World-System*.

Norihisa Yamashita is associate professor of history and cultural anthropology at Hokkaido University, Japan.

Deniz Yükseker is assistant professor of sociology at Koç University, Turkey.

[The contributors were all members of the Research Working Group on the Structures of Knowledge of the Fernand Braudel Center, Binghamton University. This book is the product of the work of the Research Working Group.]